Sustainable & regenerative materials for architecture

A source book

Sustainable & regenerative materials for architecture

A source book

Will McLean & Pete Silver

Laurence King

Published in Great Britain in 2025 by

Laurence King
An imprint of Quercus Editions Ltd
Carmelite House
50 Victoria Embankment
London EC4Y 0DZ

An Hachette UK company

A CIP catalogue record for this book is available from the
British Library

TPB ISBN 978-1-52943-327-2
ebook ISBN 978-1-52943-328-9

10 9 8 7 6 5 4 3 2 1

Commissioning editor: Liz Faber
Design: Blok Graphic
Project editor: Gaynor Sermon

Printed and bound in China by C&C Offset Printing Co., Ltd.

Contents

Foreword 07

Introduction 08

CHAPTER ONE

Materials Overview 13

A brief history of construction materials 14

Technology transfer and learning from other industries 18

Carbon (ac)counting 21

A circular economy: reduce, reuse, recycle and repeat 24

CHAPTER TWO

Extractive Materials 27

2.1 Earth & Clay 30

2.2 Concrete 50

2.3 Stone 82

2.4 Metals 86

2.5 Glass 98

2.6 Sand 110

CHAPTER THREE

Grown Materials 115

3.1 Timber & Cork 118

3.2 Paper 138

3.3 Hemp, Bamboo, Grasses & Palm 150

3.4 Seaweed 168

3.5 Mycelium 172

3.6 Algae 180

3.7 Biopolymers & Recycled Polymers 186

3.8 Wool 194

3.9 Salt 198

Endnotes 202

Bibliography 202

Glossary 203

Index 205

Credits & Acknowledgements 208

'We need to re-address our relationship with the planet by understanding what it is made of…'

Seetal Solanki, Founder and Director of Ma-tt-er

Foreword by Jonathan Smales

Founder & CEO of sustainable development company Human Nature

In 1989, when I was commissioning a new HQ for Greenpeace UK, I would have killed (or loved someone to death…) to get my hands on this book.

We were repurposing a semi-derelict building – in that instance a former animal testing laboratory – and wanted to craft a 'green' building. But how? What might that be? Which materials might we use? Why? And where would we get them from?

We found the *Good Wood Guide* by Friends of the Earth, exhumed a copy of Stewart Brand's *Whole Earth Catalog* and pored over Bucky Fuller and Victor Papanek for clues. Fortunately we had architect Peter Clegg and environmental engineer Patrick Bellew to navigate, and they did a fine job.

Now, we have this well-organized, beautifully illustrated and elegantly written guide by Will McLean and Pete Silver. It's not just useful, it's actually delicious: filled with gorgeous, organic, sensuous, natural forms. In fact, some of the natural materials they survey probably could be eaten. I may be getting carried away here.

Climate weirding is out of control. Last year, global temperatures hit the Paris target of 1.5 degrees of heating, and on current policy – still less practice – we're headed for a catastrophic 2.7 degrees. And we've embarked on what the journalist Elizabeth Kolbert and others describe as the 'sixth great extinction', erasing and degrading nature across the planet.

How did it come to this?

Here in the UK, in 2024 the new Labour government committed to build 1.5 million homes in its first term. If these houses are built using mainstream construction materials, the carbon used in production will explode the nation's carbon budget. It's a disaster waiting to happen. And it's utterly avoidable if we're brave, purposeful, strategic and wise. Designers must be centre stage if we're to succeed.

So this book is timely. It illustrates the growing wealth of choices we now have when designing and procuring buildings. The research is done or in progress, wonderful designers are showing the way forward, and we now need radical innovation in financing, building regulations and industrial strategy to bring the most promising materials and systems to the mainstream market.

Will and Pete inspire us here. They have trawled the network of excellent scientists and designers to find materials, design and construction that leap off the page and make you want to become a natural builder without delay.

At the Phoenix in Lewes, East Sussex, UK, my company, Human Nature, has designed and is preparing to build a whole neighbourhood from timber and other bio-based materials. It's not an easy gig. But it has to be done, and the growing field of work in natural building gives reassurance and reminds us of the prize, which is beauty as well as an embodied carbon account our consciences and the world can live with.

It's time to start living on the planet as if we intended to stay. In the pages that follow, you'll find part of the answer to the question, 'how?'

Introduction

Architecture and construction are at a turning point, with an environmental imperative that demands we look for new and viable material solutions. A fuller understanding of material properties allows us to make informed decisions about material choice, and this book serves as a compendium of low- and no-carbon materials (bio-based, grown, upcycled and recycled) for architecture and design. It focuses on sustainable materials, their sourcing, technical properties, and the processes required for their use in architecture.

While there are some excellent materials science books available, they can be technically dense and do not always demonstrate how these materials (distinct from product design) are used in building and construction projects. *Sustainable and Regenerative Materials for Architecture* is designed to strike a balance between analytical materials data, design intent and associated fabrication and construction processes.

There is a creative explosion of work taking place in architecture and design schools, exploring materials such as clay and earth, mycelium, engineered timber, bio-based polymers and algae. With a better understanding of the social, environmental and economic sustainability of any given material – alongside its technical properties – students of architecture can lead the change in responsible and creative material use.

As a designer, one of your most important roles is as a specifier, and the more you understand about materials and their appropriate use with respect to safety and performance (technological and experiential), the more likely it is that a project can meet specific programmatic, technical and environmental objectives. When teaching and studying architecture, it is essential that students are introduced to the world of materials and associated fabrication processes early; the physical 'stuff' of materials can both stimulate the design imagination and inform projects.

Throughout the book we showcase new and rediscovered processes for material fabrication, responsible sourcing and creative material design. Material properties (structural, thermal, fire, health and life safety) are described, and case studies help to illustrate the inventive ways in which these materials can be deployed in the built environment. More information about the fabrication, provenance and cost (both ecological and economic) of materials

RIGHT European hardwood timber, which is sawn and separated then seasoned by air drying to prevent twisting and splitting. It takes about a year for a 25mm (1 in) thick piece to season.

'Recycling is seen as the ultimate means to reboot the value potential of used materials, without the need to produce new ones. The model consists of technical cycles of mined materials and biological cycles of farmed materials. The main difference between these is that biological materials are digestible and don't require active recycling'.[1]

Adriaan Beukers and Ed van Hinte, authors of *Designing Lightness*

will also help students to navigate this rapidly expanding field. Alongside an illustrated palette of sustainable materials and exciting design precedents, the fabrication and construction technologies are described, informing students about the techniques and processes used to transform the materials into viable construction products.

Structure of the book

Chapter one provides a general introduction to the history and evolution of construction materials, while broadly setting out the terms by which material properties are described and classified. Here we examine new, novel materials and their associated fabrication and construction processes, as well as ancient local and vernacular materials and fabrication techniques that have been re-imagined for a contemporary construction environment.

This introductory chapter follows the structure of the materials classification chart and describes key criteria by which materials are selected, specified and applied. It answers questions such as: how do these materials behave structurally? How and why would you choose these materials in relation to specific programmatic requirements? What is the durability of the material over time? How easily are these materials transformed into building components?

Chapters two and three contain case studies that outline the current status of extractive (chapter two) and of grown (chapter three) materials. The construction industry is one of the largest causes of greenhouse gases, and each of the featured entries is designed to either mitigate or eliminate associated carbon emissions. In the case of regenerative materials, the sequestration of carbon (biogenic carbon) during material production and/or processing will result in a net reduction in embodied energy.

The case studies have been selected for their technical, aesthetic and social innovation, and also represent a global geographical spread of invention and entrepreneurship. Some of the materials featured are entirely or partly composed of upcycled or recycled materials, thus mitigating embodied carbon and usefully employing waste streams as part of a construction circular economy.

Without understanding our local climate context, the use of regenerative materials (however good their sustainable provenance and technical performance) is moot. If the history of architecture and construction has taught us anything, it is that the siting, orientation and form of built structures are the key determinants of the social and environmental success of a given building. Good buildings are adaptable, and the building fabric should readily respond to diurnal, seasonal (and unseasonal) climatic change. The use of sustainable and

regenerative building materials must go hand in hand with thoughtful environmental design, presenting a holistic approach that produces great architecture which is not damaging to the planet.

A note on material properties

Analytical data such as embodied energy (measured in carbon) is indicated for each featured material where possible, as well as other key technical metrics (as applicable or available) such as compressive and tensile strength, fire resistance and thermal insulative value (U-value). Information on material sourcing and associated processing and fabrication is also described to help build a bigger picture around the appropriate use and specification of each material. In some of the case studies, such as TerraCool (p.42), we see a slip-cast terracotta building component that, while using the extractive mined material of terracotta clay and associated energy of a kiln-fired process, provides a passive cooling effect and thus mitigates the operational energy of a comparable air-conditioning system. As such, the metrics of each material and its associated embodied carbon must be viewed in the context of their designed use, and it is only when utilized as part of a project that the true environmental value and cost of a material can be understood.

This book presents an overview of materials and material processes, and it also showcases the work of designers who are making it their business to better understand what the designed world is made of, where those materials come from, and how those materials perform. It might seem obvious that design decisions have consequences, but somewhere this message seemingly got lost. The rediscovery and embracing of sustainable material sourcing, making, re-making, reuse and recycling is exciting and challenging; it offers us a positive future where the crude exploitation of natural resources is replaced by a more thoughtful, circular or regenerative approach to design.

LEFT Herzog and de Meuron's Elbphilharmonie in Hamburg, Germany, uses natural gypsum plaster and paper pulp to create 'acoustically microshaped' panels in the auditorium.

'We are doing very little to adapt to the inevitable change in climate. The housing we have is not well insulated, so it leaks heat in the winter, but in the summer the natural cooling is not there as well. And the design of our urban environment is not conducive to making the higher temperatures more liveable … the design of our cities, our offices, everything we are doing is just not taking account of the changing climate.'[2]

Professor Sir Brian Hoskins, Meteorologist and Climatologist

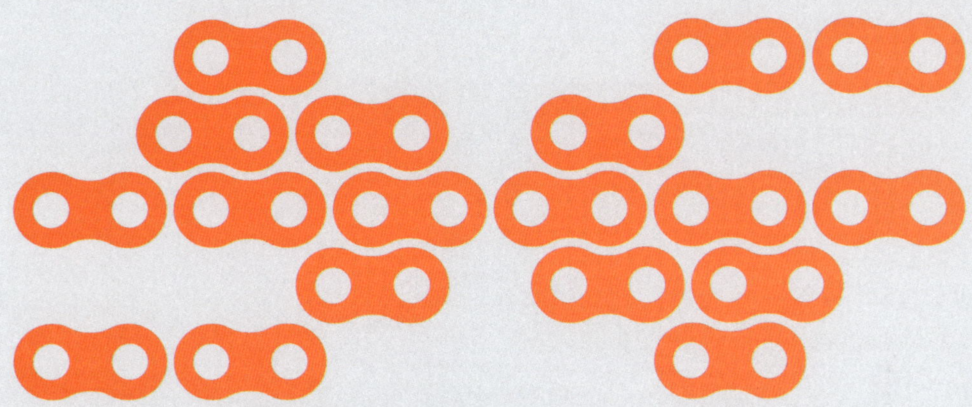

Materials Overview

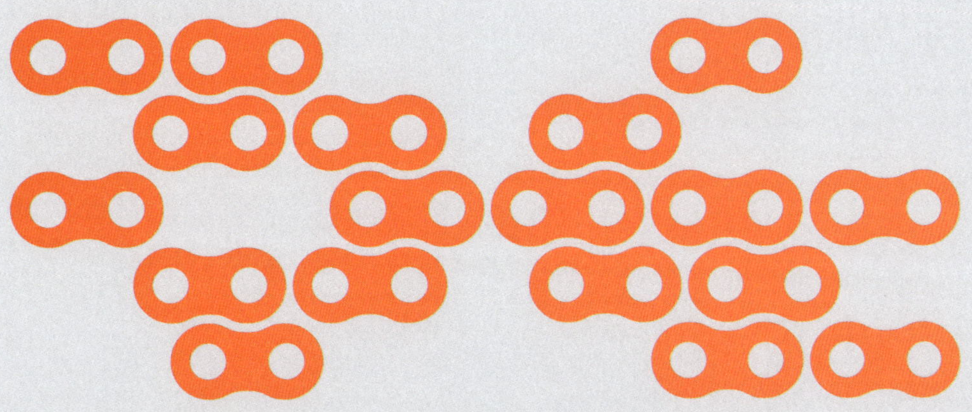

A brief history of construction materials

'The prehistoric inhabitants of the Philippines could make ropes and baskets from plant fibres almost 40,000 years ago, according to an analysis of stone tools. The find suggests the people living then may have been able to construct more sophisticated things, such as boats and buildings, than previously thought.'

Soumya Sagar, researcher and writer at *New Scientist*

Older structures are generally likely to be low tensile in nature, with materials such as earth, stone and brick used largely in compression, and are therefore unsuitable for long spans or great height. However, as the Philippine discovery shows, using sophisticated tensile structures made from plant fibres such as palm for shelter is just as old. At the same time, these ancient structures were not necessarily less environmentally efficient than contemporary ones, with passive environmental control of heating and cooling inherent in much traditional and local vernacular architecture. Traditional dwelling types were born out of the local environment, both making use of local physical resources and responding to local climate. From trees and rocks to mud and snow, each material has found a way to be employed as both structural form and climatic envelope.

Thousands of years BCE, there is evidence of mud brick construction, domes and teepees; and still in evidence today are the great granite pyramids and obelisks of ancient Egypt. The ancient Greeks began to develop an understanding of structural analysis, and the Romans began to develop sophisticated vaulted brickwork. The Medieval era saw the development of large-scale brick and stone vaults and domes in European and Middle Eastern architecture, and even of the use of wrought iron for pagoda construction in China. But it was in the period leading up to and including the Age of Enlightenment (late 17th–18th century) that there was an explosion of scientific advancements in understanding the behaviour of materials and physical structures.

The Industrial Revolution brought advances in materials science and manufacturing processes, leading to an expansion in the building of tall and long-span structures. The development of chain and wire rope, for example, not only enabled the construction of suspension bridges, but also of lifts, without which there would be no skyscrapers. Off-site manufacturing – the ability to fabricate building elements in the controlled environment of a factory or workshop – resulted in better quality control, and the capacity to transport ever larger components progressively expanded the possibilities for using prefabricated building elements.

OPPOSITE Meticulously detailed stone fan vault in the New Building, dating from the late 15th century, Peterborough Cathedral, UK.

Prior to the Industrial Revolution, structures had been designed mainly through trial and error, and usually by the craftsmen and artisans – the master builders – who constructed them. Knowledge was passed down through guilds and trade apprenticeships. Most construction technology employed either timber frames (there is evidence of triangular trusses dating back to the 6th century BCE) or low-tensile masonry (used to dramatic effect in the stone vaulting of the Renaissance and Gothic periods). From the 18th century onwards, however, materials science and materials technology, along with the application of structural engineering, offered an expanding palette of design possibilities; specifically, the principle of what Frank Lloyd Wright termed 'tenuity'[1], the ability to imbue materials and structures with tensile strength such that the capacity to span and rise, was increasing exponentially.

Although there is evidence of using metal for the bronze roof trusses of the Pantheon's portico dating as far back as 125CE[2], the first bridge to be made entirely of cast iron was completed in 1780 (the Ironbridge in Shropshire, UK, designed by Abraham Darby III and Thomas Farnolls), and also during the 1780s the roof structures of Paris's Théâtre Française and the Salon Carré in the Louvre were the first of their kind to be built entirely out of iron. In 1797, Ditherington Flax Mill in Shrewsbury, UK, designed by Charles Bage, became the first iron-framed building in the world, known as 'the grandfather of skyscrapers'.

During the 1800s, Portland cement was patented, ferro-cement was invented (for the construction of a rowing boat), and the first patent for reinforced concrete was granted (to Parisian gardener Joseph Monier, for a plant pot). Cast iron gradually replaced wrought iron, and was in turn replaced by steel (through Henry Bessemer's patent of his conversion process). By 1889, the Eiffel Tower had been constructed (still using wrought iron) by Gustave Eiffel and Maurice Koechlin, and in the same year the Forth Rail Bridge was built using steel across the Firth of Forth in the east of Scotland by Benjamin Baker, Sir John Fowler and William Arrol.

By the 20th century, it was reinforced concrete that was having an enormous influence on architecture. From the modernist architectonics of Le Corbusier's Villa Savoye (1931) and the Rietveld Schröder House (1924) to John Lautner's majestic Arango Marbrisa House overlooking Acapulco Bay (1973); from Robert Maillart's use of constant-force trusses in his Chiasso Shed (1924) to the thin-shell structures of Pier Luigi Nervi's Palazzetto dello Sport (1957), Felix Candela's Los Manantiales restaurant in Xochimilco (1958) and Heinz Isler's Wyss Garden Centre in Zuchwil (1961); from the subtle interventions of Carlo Scarpa at Museo Civico di Castelvecchio (1959–1973) to the organic, fluid forms of Eero Saarinen's TWA Terminal at JFK Airport (1962) and Oscar Niemeyer's Brasília (1956–1961): concrete came of age. Eduardo Torroja founded the International Association for Shell Structures (IASS) in 1959.

By the 21st century, innovations expanded and continue to expand the construction technology palette: the ability to create reinforced plastics – originally with glass fibres and then with synthetics such as carbon fibre and aramid (long-chain synthetic polyamides), and now with natural fibres; to produce stronger and lighter metal alloys; the technologies that have allowed glass to become a structural component in its own right; the capacity to bend an I-beam; to reinforce concrete with fibres; to engineer structural timber (through linear- and cross-lamination). On top of this, the drive for lightness combined with stiffness and strength from the industries that create mobile products – aeronautics, shipbuilding and vehicle manufacture – has created opportunities for technology transfer. Architecture continues to borrow technology from these industries.

OPPOSITE The wrought iron roof and cast iron columns of John Boucher and James Cousland's Kibble Palace. Constructed in Coulport in 1871 and moved to Glasgow Botanic Gardens in 1873.

Technology transfer and learning from other industries

'We should not be growing cotton anymore, and the use of plastic fibres in garment production should be stopped!'

Mohsin Sajid

Fashion & textiles

Mohsin Sajid is a denim specialist who has worked for some of the best-known apparel companies in the global fashion industry for over 20 years. This interest led him to look at the sourcing and manufacture of denim and its environmental and ecological effects. As a result, he is committed to replacing cotton with hemp, which is faster-growing and less labour- and chemically intensive to produce. Hemp is rougher in texture, but this is something the industry is working on. Large, branded denim companies have not used natural dye since the 1800s, and the synthetically produced indigo is environmentally problematic, so new bio-based dyes are being explored and tested. Sajid is also pioneering the use of removable metal rivets (used in jeans), as fixed rivets are currently a barrier to recycling. Sajid is also working on zero- and ultra-low-waste pattern-cutting templates. This is the kind of radical rethinking that would surely benefit the construction industry: a top to bottom rethink around sourcing, processing, fabrication, transportation and construction of materials.

Charles Ross is a textile specialist and lecturer in performance sportswear design and sustainability at the Royal College of Art. His work is focused on circularity in the apparel industry, and how to combat overconsumption with 'sustainability through longevity'. Ross describes how the fashion industry is regarded as a sustainability 'pariah' (ecologically and socially), but he also identifies that the construction industry is much worse in the production of CO_2 and waste. While the perception of 'fast fashion' as being wasteful and environmentally damaging is in many cases rightly deserved, some large retailers and producers are making big steps to improve their social and environmental sustainability. Ross highlights Timberland's 'Regen' initiative, with the leather footwear business sourcing the majority of hides from the waste stream of the US meat industry. Apparel company Patagonia remains a model of sustainable production, and while their business does use synthetic fibres, it is from recycled plastics only. Ross also commends the use of textile industry accreditation systems such as bluesign®, Fair Wear, and the Sustainable Apparel Coalition's (SAC) Higg Index, with some being better than others but all aiming to improve the sustainable sourcing and production of materials.

Dr Manel Torres launched Fabrican, a spray-on fabric technology, in 2003. His research had investigated novel ways to speed up the laborious process of constructing garments for the fashion industry. He envisaged a material that would magically fit the body like a second skin and yet have the appearance of clothing. Years of research and experimentation culminated in the realization of a sprayable fabric from an aerosol can. The technology of Fabrican is wonderfully demonstrated with a film of a 'spray-on'

T-shirt, which is then removed as a washable fabric textile[3]. Torres has created a technology that is transferable to many types of industry, and he describes medical applications (bandage and strapping) as well as use in controlling oil spills. His company successfully licensed the technology and is working with partners across a whole range of commercial applications.

Torres's product is reminiscent of technology used for the Cocoon House, Sarasota, Florida (1950), designed by Ralph Twitchell and Paul Rudolph. The roof of the Cocoon House was formed with a catenary cable and mesh armature, and then a waterproof spun-bond polymer surface was applied with the aid of a compressor and spray guns. This technology had originally been developed by the US military to wrap and protect 'mothballed' ships after World War II, and Paul Rudolph, who had served in the military for a year, had seen this industrial process and repurposed it in this interesting housing project.

Product design

Dr Artur Mausbach is an architect and urban planner who graduated from the University of São Paulo, Brazil. For his final architecture degree project, he presented a two-seater city car realized as a full-scale model. He is a research fellow at the Royal College of Art, where he is also a research supervisor and project leader in its Intelligent Mobility Design Centre. Mausbach brings experience in multidisciplinary work to collaborative research projects, addressing the challenges of human transport and mobility in the 21st century. Due to the UK's Zero Emission Vehicle (ZEV) mandate, there will soon be 15 million redundant cars in the UK. Similar mandated schemes have been pioneered by

BELOW Ecofitting a Golf. Developed by Dr Artur Mausbach at the RCA, this project embraces a circular economy strategy to retrofit a fleet of internal combustion engine vehicles to electric.

California (and nine other ZEV states), and in China, where the New-Energy Vehicle (NEV) scheme is a more stringent and nuanced credit-based system levied on volume car makers. Mausbach's 'ecofitting' project practically addresses wholesale replacement of vehicles by retrofitting vehicles with new low-cost electric motors and battery units and explores a new 'subjective sustainability'. His disruptive industry platform marries a new design paradigm which challenges the 'aesthetics of perfection' and embraces reuse, customization culture and modification.

Ehab Sayed is a sustainable designer, engineer, circular economy strategist and built environment innovator with a passion for creating a biomimetic (nature-inspired) circular economic future. After extensive research on the global construction industry, he founded research and development company Biohm to champion the integration of biological processes in manufacturing. Biohm has produced a mycelium thermal insulation panel that will be the world's first accredited mycelium insulation product, and is also developing new products and alternative applications for mycelium, which is the vegetative structure of a fungus. Sayed points to the annual 2.2 billion tonnes (2.4 billion US tons) of construction waste that is globally produced each year[4], and asks how we might begin to utilize new bio-based materials such as mycelium to not only reduce this waste, but also to actively work with current waste streams to produce low- (or no-) carbon construction stuffs. Sayed is most interested in the synergies of different parts of the construction industry, and with Biohm is challenging the status quo of material provenance and ecological and social sustainability.

BELOW The Paper Dome, a hemispherical paper tube structure designed by Shigeru Ban as a temporary theatre in 2003. Originally located in Amsterdam, it was relocated to Utrecht in 2004, where it remained until disassembly in 2012.

Carbon (ac)counting

'One major benefit of biobased materials, particularly those grown from plants, is that the plant may absorb carbon dioxide (CO_2) while growing. This sequestered CO_2, also referred to as biogenic carbon, is then trapped in the material when it is harvested.'

Arup and Material Cultures

While researching and collating material for this book, we have made our best efforts to include analytical information about the various materials and processes detailed. These numbers might include the relative compressive strength of a material, and the embodied carbon associated with the sourcing and production of a material. It is important that the numbers related to embodied energy are understood within a context, for instance whether that figure refers to biogenic carbon (any carbon that has been sequestered during the production of a material).

This might be most easily understood by looking at the production of construction timber. If timber is sustainably sourced from a managed forest, we can rightly assume that, during the growth of the tree, carbon dioxide has been absorbed (or sequestered) in that tree, and thus the timber is carbon negative. However, a single-species softwood tree plantation, while contributing to reducing carbon emissions, might not provide a particularly rich or biodiverse habitat. So while we may agree that 'wood is good', creating a monoculture is not, and like any problematic, that of de-carbonizing our environment requires nuance and a thoughtful, more comprehensive approach; in the case of UK timber production, this might include improved management of our native hardwood species as part of a regenerative construction ecology.

In schools of architecture, hence in architectural practice, we are often presented with and indeed present de facto solutions to problems. We must be wary that in some cases these problems are neatly self-serving, especially when readily answered by our pre-prepared solutions. Similarly, when new materials and new material processes are presented as the singular answer to climate change and ecological sustainability, we must maintain our critical judgement. The problem of the technological 'panacea' is that it is self-serving and necessarily discounts or omits other ideas and approaches that run contrary to the logic of this or that material or technological innovation. The key to a sustainable or regenerative approach to architecture must be multi-faceted and thoughtful. We must ask the questions: what is my building made of and why? Where can I sustainably source materials for a new construction project? What is locally available, and how might these choices support local industry, craft, skills and training?

There are huge carbon savings to be made in the construction industry to achieve the government's net zero pledge of 'committing the UK to bring all greenhouse gas emissions to net zero by 2050.'[5] The UK pledge to achieve net-zero greenhouse gas emissions by 2050 is mirrored in the EU's European Green Deal and the US Federal Sustainability Plan. The whole business of construction needs to engage in a transformation of how we build and of what we build with. This should be necessarily led by government funding, regulation and legislation. Thus far the UK government has launched the Industrial Decarbonisation Challenge, which aims to create 'one low-carbon industrial cluster by 2030 and the world's first net-zero industrial cluster by 2040'[6]. Unfortunately, in the UK, the de-carbonizing of 'hot' industries like steel and glass production might currently be seen as a useful smokescreen for de-industrialization of these complex and carbon-dirty industries. However, we must be aware of the unintended consequences of a zero-carbon environment where dirty 'high-carbon' activities are judiciously moved offshore to another regulatory domain. Ecologies are local but air pollution and other deleterious environmental effects are evidently global. To that end, we must remain locally responsible and accountable for maintaining environmental standards and good practice. What kind of achievement would UK net zero actually be if we have simply moved our carbon production offshore?

Numerical data is important as a guide to the strength, fire-resistance and carbon content of a material. The introduction of biogenic (or sequestered) carbon also begins to flesh out these numbers and provides a more comprehensive 'whole life' measure of carbon content. However, it is important to understand the larger context of specific material provenance and the future potential for that material to be reused or recycled.

The Construction Material Pyramid[7] is an interactive online tool developed by CINARK at the Royal Danish Academy, School of Architecture and Vandkunsten Architects (see p.169) that provides a useful (and visual) tool for assessing the CO_2 potential of materials for a construction project. For example, it is clear to see that aluminium sits at the top of the Global Warming Potential (GWP) pyramid, but this does not give us the whole picture of a material that is infinitely recyclable and is largely recycled. The mining and smelting of bauxite (aluminium ore) into aluminium is hugely carbon intensive, and while this can be mitigated by powering the process with zero-carbon energy, the larger question might be whether we should still be mining bauxite for aluminium? Should the extraction of planetary resources now stop, and the management of the existing resources of steel, aluminium, plastics and other mined material become about improved recycling and resource management, a process where waste is simply no longer contemplated?

LEFT The Sharp Centre for Design, Alsop Architects, Toronto, Canada, 2004, demonstrates the strength of hollow steel tubular legs and an 'inhabited' two-storey truss that forms the art school.
OPPOSITE The Montreal Biosphère, formerly the American Pavilion of Expo 67, Buckminster Fuller's lightweight dome of tubular steel.

A circular economy: reduce, reuse, recycle and repeat

'Pollution is nothing but resources we're not harvesting. We allow them to disperse because we've been ignorant of their value.'

Richard Buckminster Fuller

A circular economy is based on principles of reducing unnecessary consumption, adaptive reuse of materials and products, and efficient recycling at the end-of-life stage. This approach was described as 'cradle to cradle' by architect William McDonough and chemist Michael Braungart in their seminal book, *Cradle to Cradle: Remaking the Way We Make Things* (2002). McDonough and Braungart remind us that 'everything is a resource for something else', and that in nature 'waste' is simply someone else's food. Buckminster Fuller's quote about pollution also reminds us

that by describing something as waste, garbage or pollution, we give licence to a throwaway approach to resources that is inherently unsustainable. In a circular economy, waste streams are understood and usefully employed for the making or growing of new things – food, materials and products. The reimagining of the construction industry that is beginning to take shape, and which is illustrated in case studies throughout this book, demonstrates how the 'waste' of one project can successfully be transformed into ambitious new work. Of course, none of this is really new, and perhaps when the rampant consumer fog

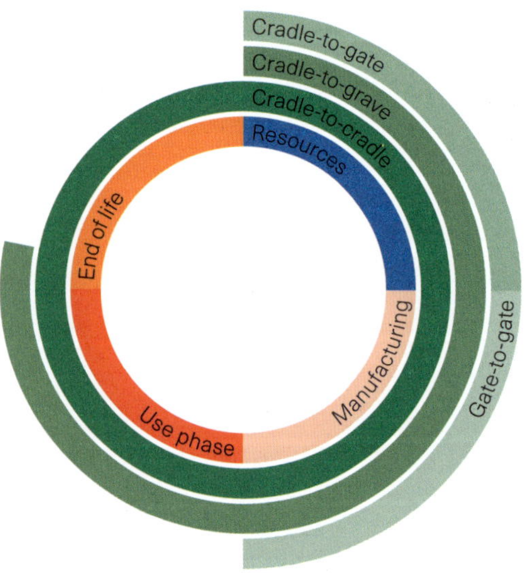

LEFT Cradle-to-cradle and cradle-to-grave Life Cycle Analyses (LCAs) cover the entire life of products, while partial LCAs, such as cradle-to-gate and gate-to-gate studies, focus on specific phases, such as materials production and product manufacturing. The diagram shown here was created by Daniel Liden.

■ **Resources**
Extraction and refinement of raw material resources into materials that can be used in products

■ **Manufacturing**
Product forming, finishing and assembly, as well as packaging and distribution

■ **Use phase**
Energy consumption and other consumables when using the product, as well as maintenance and repairs

■ **End of life**
Waste collection and processing, which in a cradle-to-grave life cycle might mean incineration or landfilling, while cradle-to-cradle options would involve product disassembly and recycling, composting or other processes that keep materials in circulation

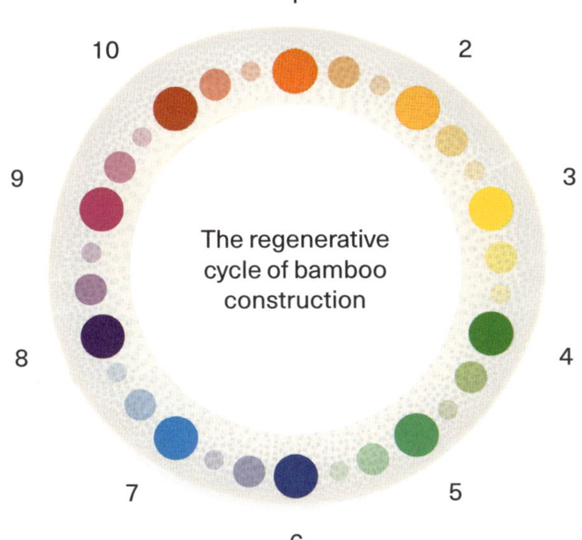

LEFT This diagram was created by natural builder Jan Balbaligo. The regenerative cycle is described in ten steps, overlaid on the cross-sectional diagram of a bamboo stem.

1. Cultivate
2. Design
3. Harvest
4. Transportation
5. Treatment and coating
6. Transportation
7. Storage and classification
8. Construction
9. Maintenance and aftercare
10. Replacement or recycling

of 'everything, everywhere, all at once' clears we might refocus on more thoughtful material choices.

Architects and designers are quick to sign up and make declarations for the good cause of environmental and ecological sustainability. However, our agency is in our actions, and what we have in our gift is the ability to specify materials and processes where we understand good provenance and minimal or positive environmental impact. While the architectural profession points fingers at fast fashion and the travel industry, the construction industry poses a much greater threat to the environment through the misuse of resources, production of unusable waste, and bad design – which poses its own health risks, such as poor daylighting, bad air quality and use of toxic materials. A more joined-up approach is required where the material sourcing for construction projects (new build or retrofit) is integral to the design. In the Bennetts Associates project for a new office and laboratory building in London's King's Cross (see p.34), architect Nikolay Shahpazov speculated on whether the earth spoil from the building foundations could be usefully employed as a construction material. Working with local brickmakers H.G. Matthews, concrete blockwork was substituted (like for like) with compressed earth (but unfired bricks), thus reducing CO_2 and using 'waste' as a resource.

Successful use of waste material is matched by the useful reuse or upcycling of architectural (or building scale) components. This is not a new phenomenon, and the art of 'spolia' is well known in continental Europe, evidenced by the disassembled columns and arches from ancient Greek and Roman constructions being usefully employed in other 'new' buildings. Steve Webb's award-winning installation 'Equanimity' at the Royal Academy of Arts in 2022 (see p.85) not only showcases the low-carbon utility of post-tensioned stone, but also incorporates upcycled stone from a recently demolished London office building.

In Studio Nauta's new housing project, De Nieuwe Veemarkt (see p.128), 130 new homes with shared community facilities are being constructed around a garden in northeast Netherlands. As an integral part of this competition-winning scheme, the construction methods and material sourcing strategy were part of the design brief; the project aims to stimulate the local economy, supporting skills and training, as well as focusing on high-quality design and the use of bio-based materials.

CHAPTER TWO

Extractive Materials

Extractive materials

Making buildings continues to use the extractive materials of earth, stone, aggregate, sand and metals. While we know that humankind has developed sophisticated enclosure systems that do not require the excavation of materials and can be made from grown timber, cultivated grasses and farmed animal skins and fur, the thoughtful use of natural and – importantly – 'local' earth and minerals remains central for any large-scale construction and infrastructure projects. In the UN report *Bend the trend: Pathways to a liveable planet as resource use spikes*, the authors explain that 'over 55 per cent of greenhouse gas emissions … up to 40 per cent of particulate matter health-related impacts … and 90 per cent of total land use-related biodiversity loss'[1] are directly linked to the ways in which materials are extracted from the earth and processed.

This chapter begins with earth-based construction, which, if used as rammed or compressed earth as described by Rowland Keable (p.40) – without the use of a cement binder – is entirely reusable and can be re-formed into new buildings for the future. Nzinga Mboup (p.31) has developed a new vernacular architecture while using the ancient technique of sun-dried earth and clay bricks. On a flight into Dakar, Senegal, Mboup had noticed how the recent concrete construction of the city no longer reflected the rich terracotta colour of the soil, which she has skilfully re-embraced. Both Urban Radicals' 'A Brick for Venice' (p.46) and Nikolay Shahpazov's straw and clay blocks (p.34) make use of 'waste' earth, re-formed into bricks and blocks and put back as usefully emblematic totems for forward-looking builders. In the earth-timber floor slab of Rematter (p.36), Tobias Bonwetsch and his team, in association with the architects Herzog & de Meuron, are pioneering a new hybrid construction system, where the earth properties of thermal mass and humidity regulation are integrated into a timber frame.

____

BELOW 'A Brick For Venice', by Urban Radicals and Local Works Studios, utilizes waste Venetian canal silt and locally recycled aggregates to produce bricks that are air dried and unfired.

'Natural resources are the basis on which all economies and societies are built, making their sustainable management critical to ending poverty and reducing inequalities. They are also essential to drive the transition to net-zero.'

Inger Andersen, Executive Director, United Nations Environment Programme

---

The section on concrete is largely split between the novel and efficient use of high-carbon cement-based materials, and the reuse or upcycling of construction waste. There is a great showcase of Concrete Canvas by Foster + Partners (p.76), the fabric formwork of Dante Bini (p.66) and the 3D-printed experiments of HANNAH in their House of Cores (p.70) and the underwater 3D-printed reefs of Enrico Dini (p.80). In Carmody Groarke's Gent Brick project (p.56) an 'upcycled waste' brick is designed, commissioned and tested as part of the scheme for a new design museum in Belgium, while the K-Briq® (p.60), designed as a low-carbon alternative to fired clay bricks, is made from construction waste and uses the existing infrastructure of the brick industry.

The contemporary use of stone as a structural material has been reinvigorated by the partnership of architect Amin Taha and structural engineer Steve Webb (p.83). With the addition of steel cabling, stone can be post-tensioned, giving it more strength than reinforced concrete while allowing for future disassembly and reuse. Stone quarrying is obviously extractive, but Webb makes a compelling argument for the abundance of (local) stone, and demonstrates its reuse in his cantilevered beam at the Royal Academy in London.

The production of steel and aluminium for construction comes at a huge environmental cost. The decarbonizing of these industries should focus on the reuse and recyclability of metals and move away from the destructive primary mining of iron ore and bauxite. The case studies focus on the designed use of metals, and their strength and engineering potential is aptly demonstrated by Schlaich Bergermann's elegant Trumpf bridge (p.90) and Jean Prouvé's (p.87) endlessly inspiring use of lightweight folded and formed steel and aluminium panels for building structure and skin. Gustav Düsing and Max Hacke remind us that lightweight tubular steel frames can provide great structural strength and material efficiency in their 'kit of parts' university study pavilion (p.94), designed for disassembly and relocation. Doris Sung's work with thermobimetals, in particular her InVert™ product (p.96), is one of the most lightweight projects featured in the book, but provides an active solar shading system without any need for environmental sensing or external power.

Glass is largely understood as recyclable, but is not as widely recycled as it should be. The Burrell Renaissance Project (p.99) shows how, with more diligent separation, construction glass can be successfully recycled, mitigating the mining of silica, whereas Faidra Oikonomopoulou and Telesilla Bristogianni's Re³ project (p.104) actively explores the various optical and structural properties of recycled glass. In Marjan van Aubel's Current Window (p.102), she explores how the use of coloured glass can be employed as part of both the design and energy system of architecture. In Sandwaves (p.111), Mamou-Mani & Studio Precht explore new types of 3D-printing technology, turning desert sand and bio-resin into a new construction material.

2.1 Earth & Clay

'I realize there is expressive potential in materials, but I'm more interested in the structural characteristics of materials than their beauty.
I think earth is the material with the most potential because it is the original source material.'

Michael Heizer, Sculptor

Earth bricks & typha reeds

**Nzinga Mboup (Worofila) and Elementerre
Location: Dakar, Senegal**

KEY DATA

Compressive strength: 2.4 N/mm^2

Fabrication process: Hand-operated mechanical press

After studying in South Africa and London, young Senegalese architect Nzinga Mboup, of architecture practice Worofila, returned to Dakar and decided to specialize in bio-climatic architecture using local materials, aiming to better understand the impact that extractive geology has on the environment. She explains: 'Vernacular architecture is the departure point for us. Whenever we study the typologies, we study the morphologies, we study the material usage; we always envisage buildings that were made to fit their context.'[2]

Working with local earth brick manufacturer Elementerre (established by Doudou Deme in 2010), special bricks were developed from crushed laterite (soil rich in iron oxide) and a small amount of sand, stabilized with maximum 8% cement. The bricks are then formed in a hand-operated mechanical press. The reason for the addition of cement is explained

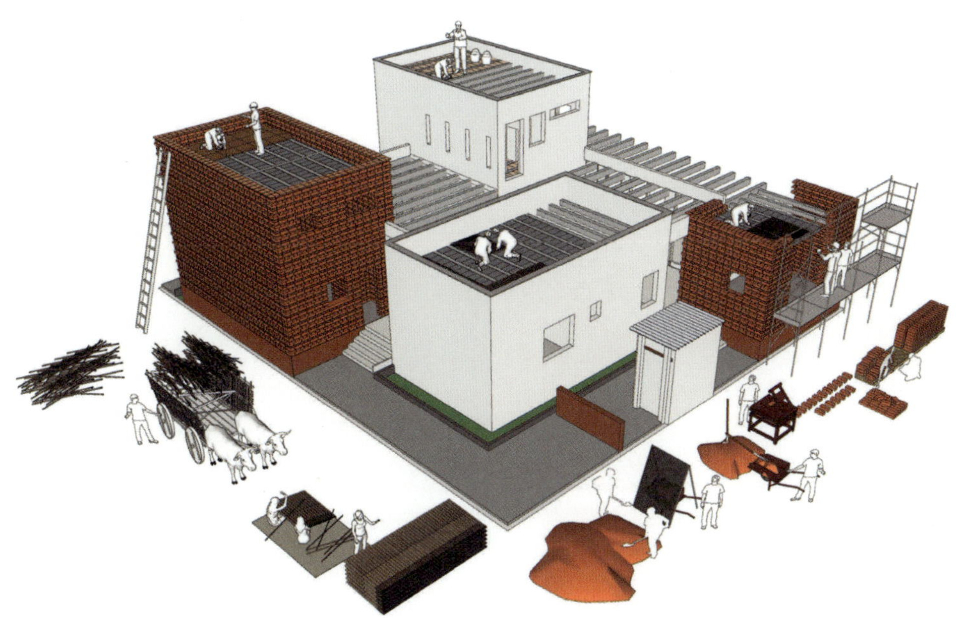

ABOVE The earth brick house under construction using the vernacular materials of air-dried earth bricks and typha reed roofing.

by the rectilinear architecture of Dakar, where walls are not sheltered from the rain with extended roof or parapet overhangs, and so the use of cement or lime helps to protect exposed walls in the rainy season. The bricks are sun-dried and not fired, but must cure for 28 days because of the added cement. The brick manufacturer has told Mboup that the use of cement has also helped to market these bricks as a durable construction material.

Dakar is situated at the western tip of Africa, and its urban landscape is predominantly concrete construction, densely planned and mostly flat roofed. Because of the geography, densification is vertical, adding to the urban heat island effect, and is increasingly dependent on air-conditioning for thermal comfort. Projects such as Mboup's multi-storey family home re-introduces the wonderfully rich colour of local earths and sands as well as utilizing their natural thermal mass. Earth walls have an excellent thermal inertia that helps to manage large diurnal temperature changes of up to 20ºC (68ºF) as well as helping to regulate internal humidity levels.

The earthen brick walls are approximately 300mm (12 in) thick with reinforced concrete foundations and ring beams at intermediate floor levels. Mboup's use of 3D modelling limits brick cutting on site, thus minimizing construction waste, and helps coordinate the positioning of structural openings and servicing. Inspired by the vernacular architecture of Asir Province in Saudi Arabia, she utilizes a horizontal banding of the exterior brickwork. This articulated ridged form of the façade provides a self-shading effect to protect the exterior walls from the sun.

BELOW The ridged form of the façade provides a self-shading effect to protect from the sun.

Because of the limited plot size in Dakar, and the fact that load-bearing earth walls for a five-storey house might have to be up to 600mm (2 ft) thick, embedded reinforced concrete columns are used to create a hybrid structure maintaining a 300mm (12 in) wall thickness. For the floors and roof structure, Mboup has used concrete beams together with infill blocks made from typha as a type of beam and block construction. She has also created load-bearing earthen floor slabs by vaulting the earth.

Typha is a type of water reed that grows on the Dakar peninsula and the northern border of Mauritania. It is a fast-growing invasive species that becomes an environmental problem when it

ABOVE The construction of the walls uses three different sizes of brick, with the two (inner and outer) leaves almost woven together and acting as a single structural surface.

blocks water flow at the mouth of rivers. There is a national programme promoting this plant as a construction material. It is an excellent thermal insulator when chopped and formed into blocks, mixed with clay and laterite for use in beam and block construction or panels, and used to line a surface on the interior or exterior. Additional benefits are that typha panels have excellent acoustic properties, and this reed does not contain any cellulose, so termites do not attack it.

'Before air-conditioning, people paid attention to materials and orientation for the natural regulation of heat.'[3]

Nzinga Mboup, Worofila

Strocks®: straw & clay compressed earth blocks

Nikolay Shahpazov (Bennetts Associates) and H.G. Matthews
Location: King's Cross, London, UK

KEY DATA

Embodied carbon: $19CO_2e/kg$

Compressive strength: $2.6N/mm^2 - 4.6N/mm^2$

Thermal conductivity: 0.2 W/mK

Fire performance: Passed an H.G. Matthews-commissioned fire test carried out by Warrington Fire

Fabrication process: Compression moulded

As part of a new development in London's King's Cross, architects Bennetts Associates were commissioned to design and oversee construction of the Apex office and laboratory building. During the groundworks phase of the project, architect Nikolay Shahpazov observed the sheer volume of earth being removed from site and speculated on whether it could be used as a construction material. After testing showed that the subsoil was free from contaminants, it was decided to divert twelve 20-tonne tipper trucks to a Buckinghamshire-based brickworks. H.G. Matthews is a small family-run handmade brickworks that has developed a range of compressed earth blocks, or Strocks®, using clay-rich earth and chopped straw for added tensile strength. The blocks are mechanically compressed and then dried but not fired, minimizing embodied energy. Strocks® are structural and have been used in projects up to three storeys high.

The King's Cross construction waste was subsequently turned into $581m^2$ (6254 sq ft) of blocks, which have been used as a non-structural element to line the basement of the building as well as provide partition walls. After a series of tests, the specially composed blocks were made from a mixture of the excavated subsoil and sand, which gives the blocks added compressive strength. The architects have calculated that by using these compressed earth blocks in place of medium-density cement blockwork, they have made a carbon saving of 84% or 9766 kg (10¾ US tons) of CO_2 emissions. The site

LEFT An unfired straw and clay brick 'Strock' made by brickmaker H.G. Matthews from earth recovered during site excavation in London's King's Cross.

'The best way to understand the difference between business as usual and regenerative architecture is by reducing the process of designing and building to its core fundamental materials… You need to question everything that you put in your (building) skin, like, where does it come from? What kind of emissions does this create? What kind of impact does this have on nature and the environment?'

Nikolay Shahpazov, Bennetts Associates

manager admitted that he'd had initial reservations about this non-standard building component, but said that installation had gone well, with the blocks laid using an earth mortar also made from the excavated subsoil.

Bennetts internal Life Cycle Assessment (LCA) calculation concluded that $19kgCO_2e/tonne$ is emitted for making Strocks®. In addition, the material does not contain any chemical binders such as cement, so at the end of the building's life the blocks can either be returned to the ground or reused for the fabrication of new earth blocks; they are also 100% recyclable, as opposed to cement blocks which can only be recycled as a low-grade aggregate. The added benefits of earth construction include excellent acoustic attenuation as well as hygroscopic properties which help to regulate internal humidity levels.

RIGHT Life cycle diagram produced by Nikolay Shahpazov showing the process of using excavated site material as a viable construction material.

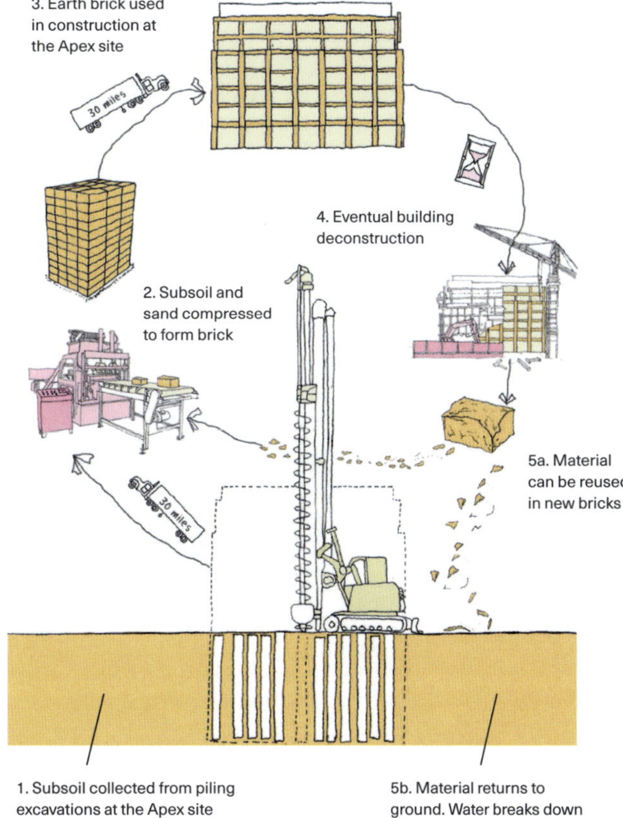

Earth-timber floor slab

Tobias Bonwetsch, Nico Ros and Remo Thalmann (Rematter), and Herzog & de Meuron Location: Basel, Switzerland

KEY DATA

Embodied carbon: (of earth floor elements) 0.27 CO_2e/kg

Load capacity: 5kN/m²

Fire performance: Certified fire resistance of RE160

Fabrication process: Compression moulded with robotic arm

Also described as a wood-and-loam floor, this construction system was devised by Rematter, a start-up company coming out of the innovation-rich ETH University, Zürich. The design team decided that, rather than develop a construction system based on a specific architectural design, they would optimize a structural framing system based on available economically and ecologically sustainable materials. The aim of Rematter is to offer a multi-storey construction system that is broadly comparable in price to that of a reinforced concrete frame, so that the larger environmental ambitions of this project are not lost through basic economics.

The primary load-bearing structure of the system is made from solid (non-engineered) spruce and pine timber columns and beams, avoiding adhesives. Rammed earth vaults sit between profiled beams, and are covered by a three-layer timber board, which helps to stiffen the floor slab and in combination with the rammed earth vaulted slabs helps to provide fire resistance. The beams are cross-connected with bolted steel tie rods resisting the lateral thrust of the earth vaults.

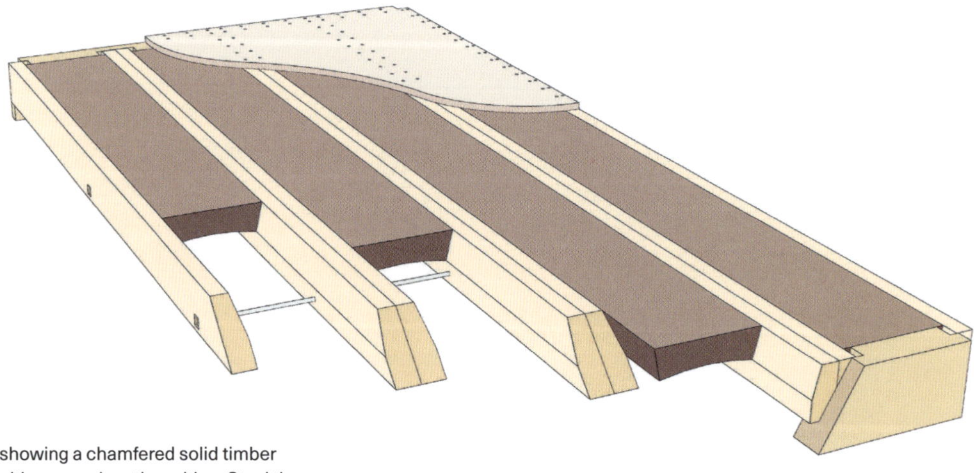

ABOVE Diagram showing a chamfered solid timber beam structure with rammed earth vaulting. Steel tie rods and a three-layer timber panel provide structural stiffness and lateral stability.

'A building that amortizes itself energetically within 30 years. Is that possible?'

Nico Ros and Remo Thalmann, Rematter

The earth for the vaulted floors is intended to come directly from site, and the subsoil excavated as part of the groundworks and foundation preparation for the building. Depending on the consistency of the soil, clay and aggregate can be added if required. However, the earth is deliberately not stabilized with lime or cement, which allows for future upcycling and reuse. The earth vaults are designed to be left as exposed ceiling soffits and add thermal mass that engineered timber slabs such as cross-laminated timber (CLT) do not have. The rammed earth also helps to regulate internal humidity levels between an optimal 40–60% humidity, and perform well acoustically, minimizing reverberation times. The earth-timber floor slab system is for residential and office construction, and designed to produce 80% less carbon emissions than a comparable reinforced concrete structure.

Working with architects Herzog & de Meuron, Rematter has utilized this floor system for the HORTUS building (House of Research, Technology, Utopia and Sustainability), a new office development near Basel. This showcase project is designed to demonstrate how the embodied energy of a contemporary commercial building can be fully mitigated over a period of less than 30 years, and how the various components and materials of the building can be reused and/or return to the earth in a fully circular design. As a part of the continued development of the system, Rematter is committed to offsite prefabrication and the fabrication of the earth flooring system is currently being automated, with robotic arms ramming the earth components to maintain structural consistency and finish quality.

BELOW Prefabrication of the vaulted rammed earth floor slabs, using robotic arms to compress the earth; finished earth-timber floor slabs for a housing project in Altendorf, Switzerland, by Jomini Zimmermann Architekten.

CobBauge

Plymouth University; Graduate School of Construction Engineers of Caen (ESITC); Regional Nature Park of the Marshes of Cotentin and Bessin; Earth Building UK and Ireland (EBUKI); University of Caen Normandy, LUSAC Laboratory; Hudson Architects

KEY DATA

Thermal conductivity: 0.26W/m²k

Fabrication process: Cast in a formwork

The project name CobBauge describes cob-earth construction in English and French, and is the name for a successful cross-channel collaboration led by Plymouth University, working with partners including Caen University, the Graduate School of Construction Engineers of Caen, Earth Building UK and Ireland, and Hudson Architects. Cob is load-bearing and made from clay-earth subsoil mixed with straw or other fibrous plant matter for reinforcement and insulation.

Cob has been successfully used as a construction material for centuries, but it no longer meets relevant building regulations without additional thermal insulation. The researchers identified that in order to meet Part L of the UK building regulations, a traditional cob wall would have to be 1m (3¼ ft) thick,

BELOW CobBauge splits the wall into two sections which are combined in formwork during construction: the inner layer is load-bearing cob, the outer layer a mix of clay and hemp shiv.

'Cob buildings are a common and attractive sight in many communities on both sides of the English Channel, but they do not meet current thermal regulations. However, with the authorities requiring new construction and renovations that are sympathetic to the historic built environment, there is definitely still a place for them. By developing new methods and training professionals in how to implement them, we can ensure this traditional technique is adapted so that it remains part of the streetscape for centuries to come.'

Professor Steve Goodhew, Plymouth University

which was seen as commercially and logistically unviable. The team felt that if they could develop a new cob wall that met regulations and was 600mm (2 ft) thick or less, then this could become a commercially viable housebuilding technology. This EU-funded research project began in 2017 and has seen material testing in the UK and France, and the recent completion of CobBauge walls in a new teaching building for the Sustainable Earth Institute at Plymouth University.

To increase the thermal insulative performance of the cob, more straw can be added, but this negatively impacts the load-bearing capacity of the material. The innovation of CobBauge is to split the wall vertically into two sections of 300mm (12 in) cob: the inner layer of traditional load-bearing cob, with an outer layer using a liquid clay/earth slip mixed with hemp shiv, which is much lighter and provides good thermal insulation. These two parts of the same wall are cleverly combined in the formwork during construction. The inner cob section is formed first, alongside a timber box void-former: when the cob is compacted, the timber box 'form' is removed and the outer layer of the wall is then filled with the hemp and clay-earth mix, with this process repeated until the wall is complete. The timber form is slightly canted, forming a zigzag cross-sectional connection (and thus better bond) between the different material densities. The general formwork for building the CobBauge walls is unusual and consists of a timber framework supporting welded mesh panels rather than timber sheets, which helps to speed drying times.

The finished composite wall is essentially cob construction of differing densities split longitudinally, with the denser inner half of the wall being structurally load-bearing and the outer half acting as thermal insulator. This new CobBauge wall has a U-value of 0.26W/m^{2}k, enough to meet Part L of current UK building regulations.

Rammed earth

Rowland Keable

KEY DATA

Embodied carbon: 38 CO_2e/kg
(calculated over a 60-year building life)

Compressive strength (rammed earth): 2–3.5 N/mm²

Compressive strength (rammed chalk): 0.8 N/mm²

Rammed earth, also known as *pisé de terre* or pisé construction, is compacted moist subsoil typically formed in a temporary formwork of the type used for in-situ concrete. Rammed earth is one of the world's oldest construction materials and was used to build sections of the Great Wall of China as well as the Alhambra in Spain. It typically consists of a mixture of subsoil, sand, gravel and between 10–20% clay – which acts as the glue binding the material together. Chalk can also be used as a rammed earth material; while not as strong as rammed earth, it does produce beautiful light-coloured walls. There are rammed earth buildings across the globe, including areas of northern France and in the UK, and one of the great benefits of this material is its hyperlocal sourcing.

Rowland Keable is the UK's leading rammed earth building specialist, a lecturer, the UNESCO Chair for Earthen Architecture and CEO of Earth Building UK and Ireland. He has led the push towards creating building standards for rammed earth that would allow the material to be more freely specified and compete against/replace the low-rise use of reinforced concrete. Rammed earth as described by Keable does not contain any cement; cement is sometimes added to the rammed earth process and is described as stabilized earth. It is important to distinguish between these materials because the addition of cement fundamentally changes this earth-based material so that it is no longer recyclable. In rammed earth, using clay as the binder forms an electrical bond between material particles, which holds the wall in place. This bond is reversible though, and a rammed earth wall can be demolished, crushed and reused as is. With stabilized earth, the addition of cement as a binder creates a non-reversible chemical bond which prevents the easy reuse

LEFT Fabricating a rammed earth wall using the same reusable steel formwork system as would be used for casting reinforced concrete. With rammed earth, the formwork rises vertically as each layer of earth is compressed utilizing hydraulic rams.

and recycling of the materials at the end of a building life. As a key proponent of earth building, Keable is keen to talk about the properties and environmental benefits of rammed earth while also acknowledging the useful structural properties of cement and concrete. But he asks the question, why are we building low-rise building structures with carbon-intensive reinforced concrete when low-carbon and cement-free rammed earth can easily do the job of structural walls for three–four storey buildings?

Rammed earth walls are typically unreinforced, 300mm (12 in) thick and have a high thermal mass, they are non-toxic and regulate the internal temperature and humidity levels within a building by naturally absorbing and re-radiating heat, coolth and moisture content. They have a similar density to concrete and are constructed by using the same formwork; with concrete a complete formwork is built and then filled with concrete, vibrated to remove air, and then left to cure for seven days before the formwork is removed. With rammed earth the formwork is part built and filled in loose layers of 150mm (6 in), which are then compacted (rammed) down to 100mm (4 in) deep with hydraulic or hand rams. When the wall has reached 600mm (2 ft) high, the next section of formwork is fitted above and the sequence repeated. When the material is compacted there is none of the hydrostatic pressure at the bottom of the formwork mould that occurs with liquid concrete, and as such the formwork can be removed as soon as the ramming process is complete. The removal of the formwork also helps with the drying process of the earth, which hardens the material.

———————————————

BELOW Unlike reinforced concrete, which requires curing time, formwork can be immediately removed after compressing the earth. This is a temporary pavilion for the Chelsea Flower Show.

TerraCool: slip-cast clay faience

Lachlan Fahy and Dilara Temel

KEY DATA

Compressive strength: 50 N/mm²

Fire performance: Fire-proof/Euro-class A1 fire-rated

Fabrication process: Slip-cast fired clay

Thermal performance: Cooling effect 12°C (53.6°F) temperature drop

TerraCool was designed in response to the increasing problem of the urban heat island and the urban heat island effect. An urban heat island can be defined as an urban area that is significantly warmer than its surrounding suburban or rural areas. This effect is more profound at night, and is caused by several factors that include waste heat generated by mechanical ventilation systems and vehicular traffic, and the thermal mass of hard urban surfaces such as stone, brick, concrete and asphalt, which store up solar energy throughout the day. The urban heat island is exacerbated by a lack of cross-ventilation in dense urban areas as well as the relative lack of greenery such as trees, bushes and grass that provides natural cooling through the transpiration of their leaves.

Evaporative cooling is a cooling process caused by the adiabatic (taking place without heat gain or loss) evaporation of water. In a process that has been harnessed by humankind for centuries, ceramic water-filled vessels can be used for local air-cooling as well as cooling water. Fired but unglazed terracotta clay vessels are filled with water, which saturates the clay and causes the outer surface to perspire (or sweat), creating a natural cooling effect. This process is similar to how human sweating helps to regulate body temperature.

Inspired by Dr Rosa Schiano-Phan's research on contemporary approaches to natural cooling, designers Lachlan Fahy and Dilara Temel have developed TerraCool, which combines the natural cooling properties of clay and the elegant geometry of the minimal 'Schwartz P' surface originally described by Hermann Schwartz in the 1880s. Minimal surfaces are being increasingly studied for use as heat exchangers, where a large surface area can be contained within a small physical boundary. This large surface area maximizes contact with the air

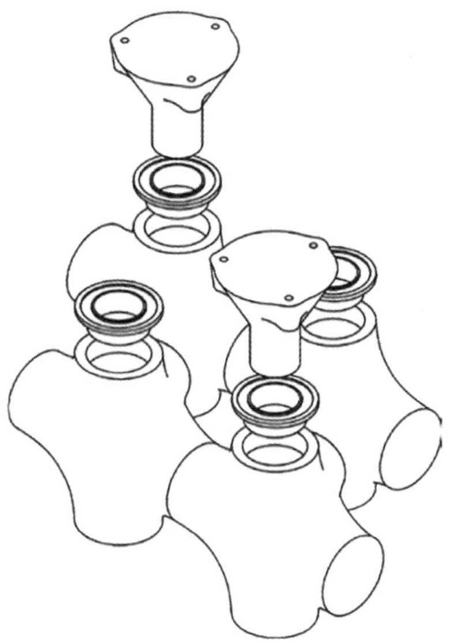

ABOVE Diagram showing the modular components of the TerraCool system.

Preparing the mould.

Creating the plaster 'negative' mould for slip casting.

3D CNC milling of a component.

Completed elements joined together to form a panel.

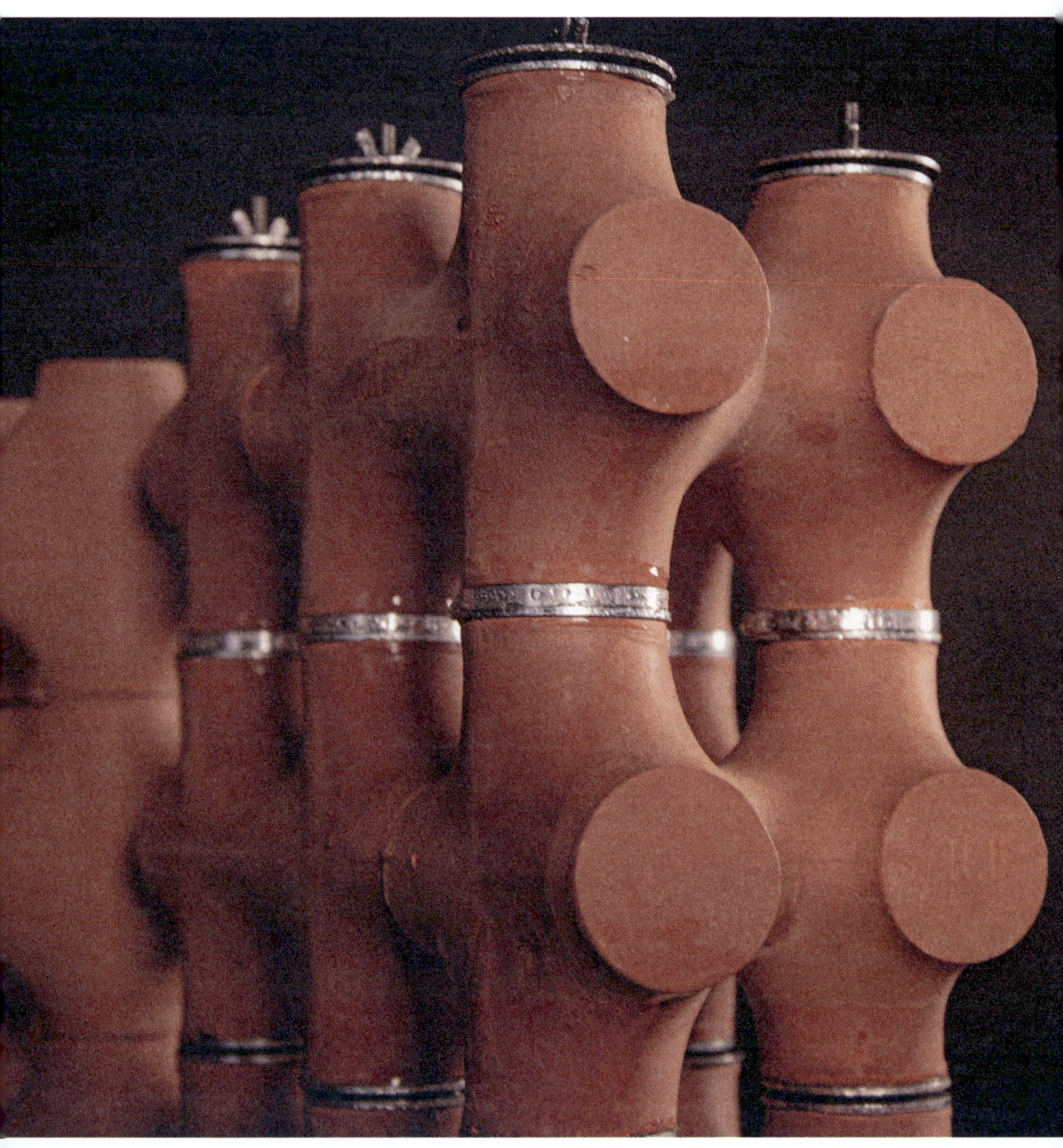

'TerraCool aims to combat the impacts of extreme heat in urban areas and promote the widespread adoption of ceramic evaporative cooling, harnessing the natural properties of terracotta to create comfortable thermal environments while reducing energy consumption.'

Lachlan Fahy and Dilara Temel

and thus maximizes the potential for local cooling. Fahy and Temel have invented a modular system of repeating hollow terracotta clay components that can be joined together in different configurations to create cooling surfaces that might wrap a building or form internal walls or thermal barriers.

The individual components are made using a liquid clay slip poured into plaster moulds, a process widely used in the fabrication of sanitaryware and faience façades. The manufacture of these complex moulds has been radically simplified by the use of 3D-printed or milled 'positives'. The rapid prototyping of differently sized and scaled components has allowed the researchers to iteratively develop this technology and subsequently model and analyse its effect using computational fluid dynamic (CFD) modelling and thermal imaging to verify the cooling potential. In addition to the design and fabrication of working prototypes of their evaporative cooling device, Fahy and Temel have also looked at how local wind flows can be studied and utilized as part of a larger project to help mitigate and temper the urban heat island effect.

OPPOSITE Finished modular terracotta elements can be joined together in different configurations. The visible condensation on the metal joints demonstrates the evaporative cooling effect.

Variable buoyancy airflow strategies

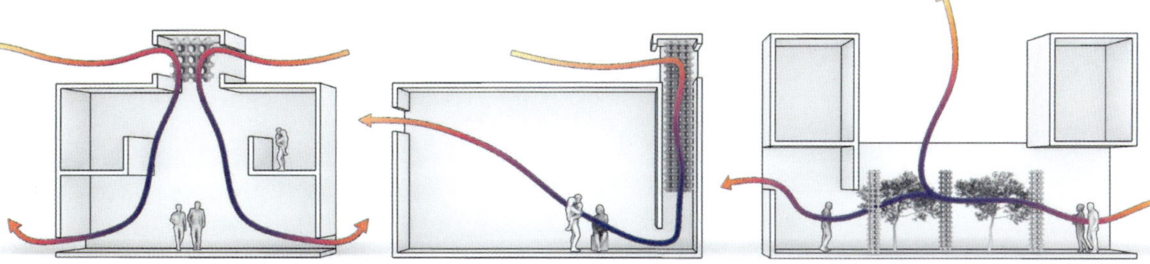

1. TerraCool bricks at the top of an atrium induce a downdraught of cool air, displacing warmer air

2. Openings at the top of the outside wall or tower allow air to enter the cavity, where it is cooled by the brick surfaces

3. The connection of adjacent spaces with different air temperatures drives cool air into the warmer space

'A Brick for Venice': silt brick

**Urban Radicals, AKT II and Local Works Studio
Location: Venice, Italy**

KEY DATA

Embodied carbon: 0.18 kg CO_2eq
per kg (estimated)

Compressive strength: 12.7 N/mm^2

Fabrication process: Compression
moulded

The ecological fragility of Venice is well-documented, with rising sea levels and building subsidence. The severity and regularity of flooding is exacerbated by the clay-like silt that builds up in the canals and which is regularly dredged and then subsequently dumped on the uninhabited island of Tresse. Produced for the Venice Biennale 2023, 'A Brick for Venice' was conceived as a semi-polemical project which realizes a technically viable construction product made from waste silt. The design team had 1m^3 (35 cubic ft) of Tresse silt and used lime

BELOW Silt from the Venice canals together with local recycled aggregates is formed in a mould, compressed and air dried.

mortar as the main binding medium. For additional compressive strength local recycled aggregates were used, including crushed concrete, brick and stone paving. For additional tensile strength, hemp fibre was added, a material that has been grown and processed to produce ropes and sails for centuries throughout the Veneto region. The bricks are air-dried and unfired, massively reducing the embodied carbon associated with conventional brickwork.

To establish the viability of the silt brick, initial proto-typing and development took place in collaboration with Sussex-based Local Works Studio, a design practice focusing on the sustainable use of local resources. A brackish silt from a Sussex river estuary was used in place of Venetian silt for the prototype.

Inspired by the funnel-shaped chimneys typical of Venice, the 'Brick for Venice' installation created a small pavilion made from a series of brick prototypes, with local masons skilfully blending silt bricks, waste aggregate and more conventionally fired clay bricks. Working in close partnership with engineers AKT II, the design of the pavilion also used surface geome-try and a combination of three different brick shapes and sizes to maximize structural strength. 'A Brick for Venice' is a proof of concept that embraces a new approach to a circular economy that finds value in abundant local waste streams.

ABOVE Temporary pavilion incorporating the new bricks, exhibited at the 2023 Venice Architecture Biennale.

'When you remove the layers of the unknown, what in the first instance might seem unachievable can in fact turn out to be a great idea. Waste silt, waste aggregates, hemp and lime … and we made a new brick – for Venice!'

Edoardo Tibuzzi, Design Director, AKT II

3D-printed Earth House (TECLA)

**Mario Cucinella Architects (MCA) and
World's Advanced Saving Project (WASP)
Location: Ravenna, Italy**

KEY DATA

Embodied carbon: 60m³ of earth construction for an average consumption of less than 6kW of electrical energy.

Compressive strength: 2.32 N/mm²

Fire performance: Naturally non-combustible

Fabrication process: 3D printed

TECLA, or Technology and Clay, is the first ecologically sustainable housing prototype 3D-printed from earth. TECLA is a partnership between Mario Cucinella Architects and the World's Advanced Saving Project (WASP). WASP was established in 2012 by Massimo Moretti, with the name inspired by the Potter Wasp, a type of wasp that builds intricate pot-like nests from earth.

TECLA's 3D-printed earth house is a small building of 60m² (646 sq ft) and consists of a peanut-shaped plan of two interconnected circles. Like the fabrication of a traditional clay coil pot, layer upon layer of 'printed' earth are piped to create a 3D clay vessel. The layers of earth extrusions are 12mm (½ in) thick, with the building 'printed' in a total of 350 layers. The walls vary in thickness from the base (600mm [2 ft]) to the crown (490mm [19¼ in]) and are made up of sinusoidal curved webbing that connects the inner and outer surfaces, creating a hollow cellular wall section that is in part filled with rice husk

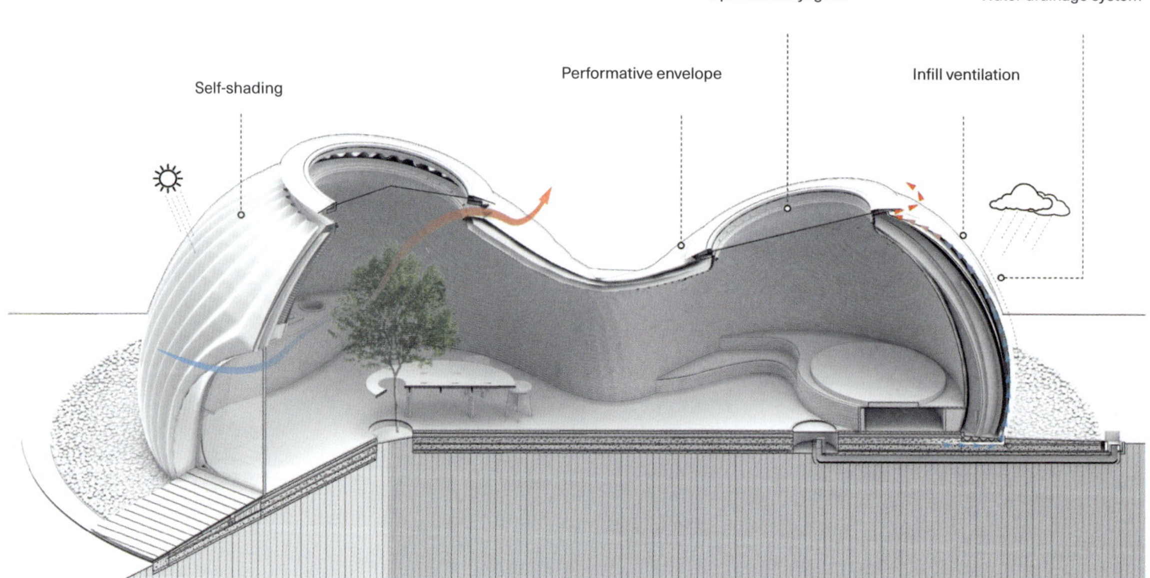

ABOVE The first eco-sustainable, low-carbon housing prototype, printed from raw earth and built in the WASP technology park.
OPPOSITE Cutaway drawing shows key environmental strategies including ventilation, water management and thermal comfort.

waste for thermal insulation. The outer form of the building is undulating to provide self-shading and to channel rainwater. The building form has been optimized for the climatic conditions in Ravenna, as have the surface morphology and the cavity dimensions and filling, with the architects explaining that all these factors can be adjusted for different climates. One size does not fit all, and we should usefully embrace the mass-customization that these 3D printing technologies afford.

The earth for the building is excavated on site, with the whole building, including foundations, using an estimated 60m² (646 sq ft) of natural materials.

The building simultaneously used two synchronized printer arms, and was printed in 200 hours over a few days, allowing for close monitoring of print consistency and quality.

Cucinella explains that this project is experimental, and is intended as research into the mass production of an ecologically sustainable house using an appropriate mixture of high technology, bioclimatic design and local, natural materials. And while these intriguing structures might not be immediately popping up in your neighbourhood, the success of the cellular printed earth walls is great transferable technology that designers should willingly embrace. As an antidote to the zipped-up logic of Passivhaus, the whole building energy efficiency certification developed by the Passivhaus Institute in Germany, Cucinella poses the question: to what extent can we custom print a variable, thermally responsive and breathable building substrate, and what can we learn from such intriguing prototypes?

2.2 Concrete: recycled, upcycled & cement replacements

'Housebuilding is critical for society and up to 300,000 new homes will be needed per year in the UK, so we need materials to plug that gap and also satisfy EU regulations around waste reduction: 70% of all building materials will need to be recycled and zero to go to landfill.'

Dr Samuel Chapman, Heriot–Watt University, Edinburgh

Supplementary cementitious material (SCM)

In their 2018 report[4] for Chatham House, The Royal Institute of International Affairs, Lehne and Preston set out the challenges for the decarbonization of the cement industry, with a particular focus on the UK cement and concrete sector. Some straightforward measures, such as improved energy efficiency of cement plants and replacing their fossil fuel energy supplies with no-carbon alternatives, are a useful start, but do not address the 50% of carbon emissions produced by the manufacture of cement clinker – the largest single ingredient of Portland cement. The production of cement clinker requires heating limestone (calcium carbonate), clay and shale in rotary kilns to 1450°C (2642°F). This creates calcium oxide, and the oxidization process releases huge amounts of carbon dioxide that had been sequestered in the seashell makeup of limestone.

BELOW Cement-free concrete blocks made with CarbiCrete technology, which enables the production of carbon-negative concrete.

'Cement is a key input into concrete, the most widely used construction material in the world. Every year, more than 4 billion tonnes of cement are produced. The chemical and thermal combustion processes involved in the production of cement are a major source of CO_2 emissions, contributing around 8 per cent of annual global CO_2 emissions.'

Johanna Lehne and Felix Preston, Royal Institute of International Affairs

---

To mitigate these carbon emissions, supplementary cementitious materials (SCMs), also known as cement replacements, can be used. Well known SCMs include pulverized fuel ash (PFA or fly ash), ground granulated blast-furnace slag (GGBS) and silica fume. The problem with these supplementary materials is that they are often the products of 'hot' carbon-intensive processes such as coal-fired power stations and furnaces for the production of steel and glass. Therefore, in a de-carbonized industrial environment, the sources and supply of these SCMs is time limited. The use of SCM in concrete is not new, and materials like GGBS have been used in large-scale infrastructure projects, such as dams for hydroelectric plants in the Scottish highlands, since the 1950s. A much older use of SCM, which predates the invention and use of Portland cement, is the use of natural pozzolanic materials such as volcanic rock or pumice which are oxidized by the heat of geological activity and share similar cementitious properties that chemically bond aggregate and sand to form concrete. Perhaps the most recognized use of pozzolans is in the unreinforced concrete roof of Rome's Pantheon building.

In addition to the established SCM technologies of GGBS, PFA and pozzolans there are also an increasing range of novel cements that not only limit CO_2 emissions but also create very low carbon or even no-carbon cements. These novel cements are designed to be produced with much less carbon and, in some cases, the curing process actually sequesters carbon. Carbonatable Calcium Silicate Clinkers (CCSC) can reduce associated production emissions by over 40% and Magnesium Oxides from Magnesium Silicates (MOMS) can even produce carbon negative concrete. Carbon negative concrete captures more CO_2 than is produced during manufacture because the concrete is hardened by carbonation using CO_2, as opposed to hydration using water, absorbing CO_2 as it cures and hardens.

While the cement sector needs to de-carbonize its production, at the same time demands for these products are rapidly increasing across the globe. So, although changing standards, technical innovations and best practices might make the industry less carbon intensive, if demand continues to increase, these improvements will be nullified. It is apparent that a different attitude to the specification and ubiquity of concrete use needs to be promoted.

OPPOSITE, ABOVE Precast concrete sections for the second Severn Crossing between Wales and England. Where reinforced concrete is required for infrastructural projects, SCMs can reduce embodied carbon.

LEFT The Donghai Bridge, Shanghai, Lin Yuanpei, 2005. The 32.5km (20 mile) bridge uses precast concrete girders and is designed to last for 100 years.

Cleancrete: cement-free concrete

Gnanli Landrou (Oxara)

KEY DATA

Cleanbrick Loko: 295 x 140 x 90mm
(11½ x 5½ x 3½ in)

Density: 2100 kg/m³

Compressive strength: 10 N/mm²

Thermal conductivity: 0.79 W/mK

Embodied carbon: 0.029kg CO_2e/kg

Gnanli Landrou is a materials scientist and entrepreneur. As a child he travelled with his uncle in West Africa, watching him build earth houses, and later researched the challenges facing the global construction industry while studying in France: energy-intensive production of cement, dwindling supplies of construction-grade sand and gravel, and the high cost of concrete – unaffordable in many countries. Landrou had experienced the benefits of building with clay/earth, but also understood how labour-intensive and time-consuming traditional clay brick construction was. At ETH Zürich, he worked with mentor Professor Guillaume Habert to develop a process to turn clay-based excavation material

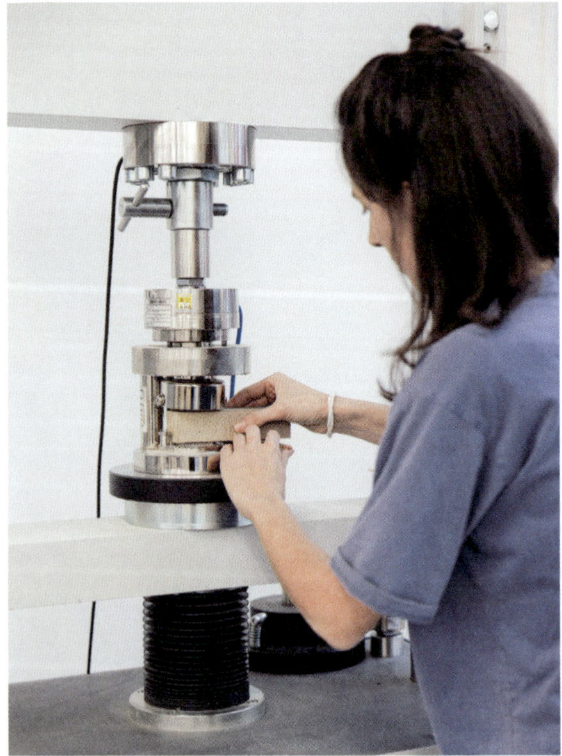

into an alternative cement-free concrete. Landrou founded Oxara in 2019 together with fellow PhD student at ETH Zürich, Thibault Demoulin.

Oxara-based building materials are 100% cement free and unfired, which reduces the material-related carbon emissions by 80–90% compared to conventional construction materials such as fired bricks or reinforced concrete. The reuse of construction 'waste' is also a key part of Oxara's mission to create a family of low-carbon construction materials and embrace circular economy principles in the high-value reuse of seemingly low-value construction waste. Oxara states that its products can contain up to 45% of materials previously designated as waste.

Oxara's technology is a patented mix of non-toxic mineral salts used as a concrete admixture and binder. This can be used for clay-based bricks and blocks as well as a cement replacement material for concrete. Added to earth subsoil and water, this special Oxara admixture enables earth to be turned into a slurry that can be poured into a formwork mould in the same way as concrete. Oxara's innovation means that the existing infrastructure of the concrete

industry, such as concrete mixers and formwork, can be utilized for the construction process. The cement-free 'cleancrete' of Oxara is only one-sixth as strong as concrete, but it could be used for two- to three-storey housing and has the benefits of rammed earth, including excellent thermal mass and humidity control without the laborious construction process. Landrou looks for ways in which we can modernize earth technology to make affordable housing with 90% less CO_2 than the cement-based alternative.

LEFT Unfired blocks suitable for use in a variety of applications, such as structural and non-structural walls, floors and cast elements (both interior and exterior).
OPPOSITE Filling moulds with a cement-free admixture and soil slurry; strength-testing a finished sample brick in the lab.
BELOW Oxara's concrete used for a load-bearing wall.

Gent Waste Brick: recycled crushed concrete & glass bricks

Sogent, Design Museum Gent, BC Materials, Local Works Studio, ATAMA Architects, Carmody Groarke and RE-ST
Location: Gent, Belgium

KEY DATA

Embodied carbon: 0.17kg CO_2e/kg

Compressive strength: 37.6 N/mm²

Fire performance: No fire testing required as the bricks are composed of inert materials

Fabrication process: Mechanically compressed

In 2019 ATAMA, RE-ST and Carmody Groarke won the competition to extend Gent's Design Museum with a new wing (DING: Design in Gent) and also to sensitively restore the three existing buildings on the museum masterplan. In order to meet the client's requirement of significantly reducing the embodied carbon of the new extension while embracing the *raison d'etre* of the museum, the design team developed the Gent Waste brick, made from 63% recycled waste of crushed concrete and glass from the city of Gent. The bricks are unfired, and estimated to contain 33% of the embodied carbon of a typical Belgian clay brick over a 60-year life cycle.[5]

To reflect the light-coloured civic architecture of Gent, the pale-coloured facing bricks and white mortar are fabricated from locally sourced municipal

OPPOSITE Waste glass excavated from the city of Gent.
ABOVE Pouring crushed waste concrete into the compressed earth block machine at BC Materials.
RIGHT Brick and block production at BC Materials.

waste streams, which include crushed concrete and white glass with hydraulic lime as the primary binding agent. The hydraulic lime also absorbs CO_2 from the atmosphere as the bricks cure, acting to sequester carbon over the lifetime of the building. Unlike fired clay bricks, these cured bricks gain strength through carbonation with atmospheric CO_2. These bricks are chemically similar to calcium-silicate blocks, but the use of recycled (silicate) aggregates mitigates embodied energy, as does the use of air-drying rather than an energy-intensive autoclave process. The recycled 'waste' silicates have pozzolanic properties which help to bind the brick ingredients.

In an article for *Architects' Journal*,[6] architects Sian Ricketts and Neil Michels of Carmody Groarke identify that the history of bricks and brickmaking was historically a local operation, with material excavated in situ, shaped and fired. With the globalization of construction products this is currently less typical, and one of the environmental benefits of this project is what they describe as 'hyper localized construction', with the waste material sourced within 7km (approximately 4½ miles) of the project and the bricks manufactured on the outskirts of the city.

Initial prototyping of the brick was undertaken in the UK by low-carbon construction experts Local Works Studio, based in Sussex, with subsequent production

'By creating a brick made from recycled waste, the Design Museum Gent is not opting for the easy solution, but rather the most ecological, innovative and visually appealing…'

Sami Souguir, Alderman for Culture, Gent

developed by Brussels-based BC Materials, who specialize in innovative earth-based construction. The bricks took two years to thoroughly develop, test and certify for use. Testing included compressive strength and freeze-thaw tests. The certification process was complex, with the design team acknowledging that current regulatory frameworks do not cope well with innovation and there was no suitable European standard to accommodate

the Gent Waste Brick. The brick was eventually certified for use in collaboration with the Belgian Union for Technical Approval in Construction (BUTGB), and the production of the 100,000 bricks is now completed.

The architects identified the need for a different approach to architectural detailing that is more sympathetic to new types of building materials, stating: 'We need to develop a new sustainable vernacular and adopt a practice of 'softer' building, whereby less environmentally onerous materials are coupled with more forgiving detailing.'[7]

OPPOSITE Front façade of the new wing of the Design Museum Gent, the facing bricks blending with the rest of the Drabstaat's light-coloured architecture.

BELOW The bricks cure and harden through a process of carbonation while air drying. This massively reduces the carbon emissions produced, compared to conventional fired clay bricks.

K-Briq®: recycled construction waste bricks

Kenoteq

KEY DATA

Embodied carbon: 16.80g CO_2e per brick (traditional clay brick = 455g CO_2e per brick)

Compressive strength: 29.0 N/mm²

Fabrication process: Mechanically compressed

BELOW The K-Briq® can be produced in a range of colours, making it attactive to designers and helping the design and construction industries to meet their net zero targets.

The K-Briq® was developed by Professor Gabriela Medero and Dr Samuel Chapman from Edinburgh's Heriot-Watt University. The product was born out of several key factors, including a UK brick shortage, a future house-building programme and the new legislation around construction waste.

The K-Briq® is made from almost 100% demolition and construction waste. Working with partners Hamilton Waste & Recycling, the waste is delivered to Kenoteq as Scottish Environmental Protection Agency (SEPA) certified inert waste. This waste is then ground and mixed with waste plasterboard, water and other 'secret' ingredients to form a thick slurry that is then compression moulded into brick modules. This circular economy model allows for the rapid transformation of a waste material into

'The K-Briq is responding to three big challenges in construction, and indeed society. Mainly brick shortages. 2.6 billion bricks were used in the UK in 2019 and over 500 million are imported, which is stifling growth and supply and projects.'

Dr Samuel Chapman, Heriot-Watt University

a saleable and robust building product. Unlike a conventional clay brick the K-Briq® is not fired, thus saving all the associated energy and carbon emissions of baking clay at 1200°C (2192°F).

The K-Briq® is cement-free and the recipe and process are patented in the UK and US. Because the bricks do not require firing, construction waste can be processed into a usable and low-carbon building product within 24 hours. Other benefits include improved thermal insulative properties in comparison to clay brick, and a range of recycled pigments can be added to vary the colour of the product. Because of the waste material provenance of the K-Briq®, it can be manufactured anywhere, ideally close to a construction waste facility, promoting localized manufacture and reducing the need for carbon-intensive transportation.

According to Kenoteq, the UK is the largest single brick market in Europe, with over 2.6 billion bricks used in 2019, with 500 million of these bricks being imported due to domestic supply shortage. In 2018 an estimated 85% of bricks used in Scottish new builds were imported because of local supply issues. According to DEFRA (the UK's Department for Environment, Food & Rural Affairs), in 2018 construction, demolition and excavation were responsible for 62% of the UK's total waste, and EU legislation now requires 70% of all construction and demolition waste to be recycled, with 0% going to landfill. With the implementation of the new UK Environmental Act 2021, the construction industry needs to rethink how it understands and harnesses different waste streams.

The K-Briq® has an estimated 4% of the carbon compared with a typical clay brick, with comparable strength and durability. With BBA certification due at any time, the K-Briq® will soon be available for specification and commercial use, with an EPD certificate to follow.

———————

BELOW Unfired K-Briqs® are manufactured using the same plants and machinery as for traditional fired clay bricks. They are produced in cooperation with local waste facilities and harness previously unused or underused waste streams.

Foredown recycled tiles

BanfieldWood

Architect Simon Banfield and his structural engineer business partner Matthew Wood were designing a new house extension in Sussex and decided to directly engage with some of the construction work themselves. Following partial demolition of an existing structure and site excavation there was an estimated 8–10 tonnes (8¾–11 US tons) of brick, concrete and other hardcore waste left on the site, which was either going to have to be removed or, they asked themselves, could they reuse some of it?

They decided to create their own tiles for cladding the vertical surfaces of the new extension. Initially conceived to have a terrazzo-type finish, tests showed that each tile would have to be polished to reveal the aggregate and, after calculating that they required over 1000 tiles, time and money meant this was not going to happen. However, there are still plans to fabricate some larger tiles for the upper section of the wall.

To produce a fine aggregate for the tiles, the construction waste was processed with a hired brick crusher (successful but loud) and reduced

BELOW The construction waste mixture is placed into vacuum-formed moulds.
OPPOSITE The architect, contractor and client all join in to help produce the tiles.

into 10–20mm (⅜–¾ in) pieces, which were also used as an alternative to Type 1 aggregate for smaller ground-bearing concrete slabs. While the hire of a brick crusher cost money, this expense was offset by not having to pay for the disposal of the construction waste.

BanfieldWood wanted a bullnose or chamfered edge to the tiles, which meant they needed to be moulded. Timber formers, or 'positives', were produced and then vacuum formed with high impact polystyrene (HIPS) plastic sheets to create moulds that could typically be used approximately 15 times. Many prototypes were produced in different thicknesses and with moulded 'lugs'

for vertical hanging, but the flat tiles of 15mm (⅝ in) thickness and moulded and profiled edges were most successful.

The casting/moulding of the tiles was done in batches of 50 tiles at a time; moulds were attached to a timber board and the waste mix poured in, smoothed, and shaken for 10 minutes to release air bubbles. As the plastic moulds were so smooth, no release agents were required and the tiles were successfully removed from the moulds within 2–3

BELOW Elevation and detail drawings showing the tile layout for these vertically hung recycled wall tiles.

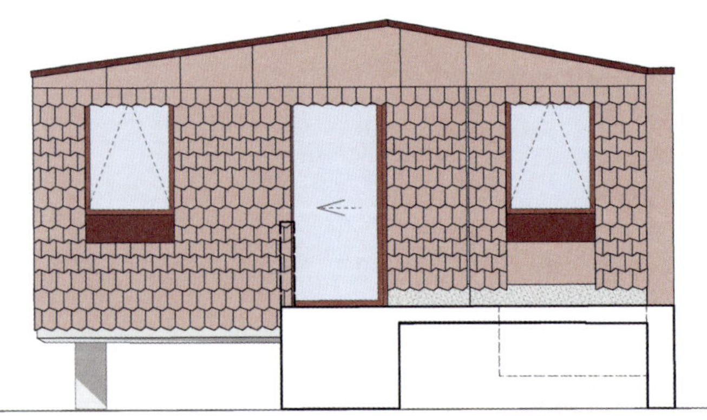

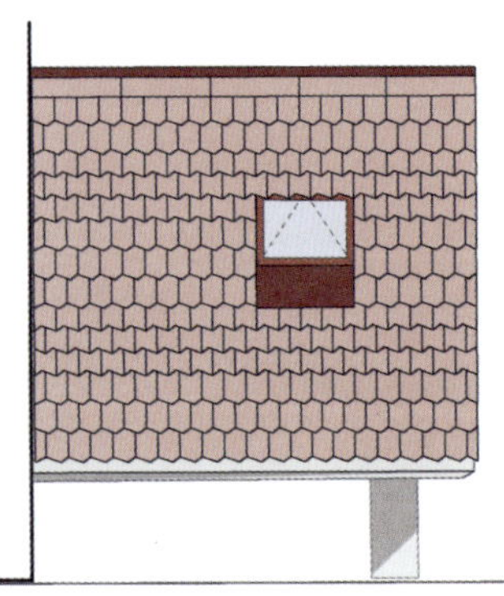

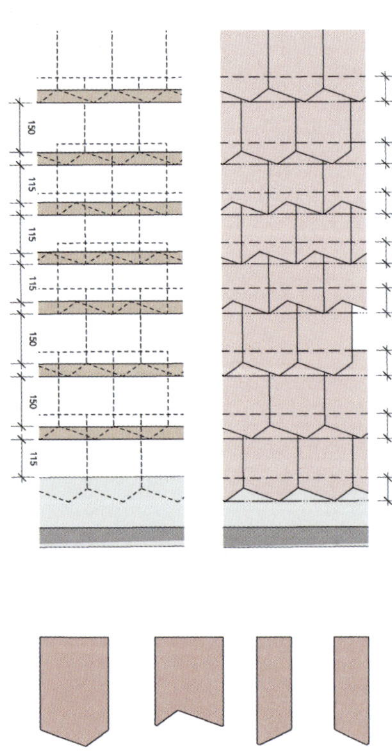

days. The tile mixture consisted of 1 part glass fibre strands, 2 parts cement, 1 part waste aggregate and 1 part building sand.

To meet the colour requirements of the planners, BanfieldWood used a metal oxide pigment additive in the tile mix to provide the correct colour palette and UV resistance. The designers found that the colour varied based on temperature and drying time, and as such there is a colour variation across the surface, which has turned out to be very successful. The reuse and craft aspect of the whole process engaged both designers and client, so much so that the client also got actively involved in the manufacture of the tiles.

'Originally, these needed to be clay tiles, but the cost here seemed fairly high, and the option of making a recycled concrete reinforced tile from this spoil became attractive.'

Simon Banfield, BanfieldWood

BELOW Finished tiles and half tiles made from the crushed brick waste.

Fabric formwork

Dante Bini
Location: Various

KEY DATA

Fabrication process: Reusable
pneumatic formwork

____

BELOW Smoothing liquid concrete on top of expandable steel
reinforcements and reusable inflatable fabric membrane. Air
pressure lifts and forms the Binishell until the concrete sets.

Dante Bini is an Italian architect, inventor and
entrepreneur. One of his earliest inventions was
bottle packaging for his family wine business, which
provided a safe means of handling 12 bottles while
boldly displaying corporate provenance and saving
on materials and cost. Emblematic of his subsequent
approach to building, for decades Bini has contin-
ually challenged the construction industry to cut
down on waste materials, innovate or automate the
construction process, and systematize a construction method for ease of repetition.

In 1964, near Bologna, Bini successfully constructed
a 12m (40 ft) diameter, 6m (20 ft) high hemispherical
concrete shell structure in three hours, using the
unique pneumatic formwork of a giant inflatable bal-
loon. Bini named the product Binishell, and over 1500
Binishells were constructed across the world from
1964 to 1990, with diameters of between 12 and 36m

'The main characteristic of this technique is that the concrete is poured on a membrane in flat position on the ground and is then lifted and shaped by inflating the membrane.'[8]

Dante Bini

(40 and 118 ft) and with a varying elliptical section. Dante Bini has acknowledged the influence on his work of the thin-shell concrete domes by architects such as Adalberto Libera and Pier Luigi Nervi, but by using inflatable fabric formwork (which he called pneumoform), he obviates the need for complex and expensive single-use timber formwork. In the 1940s, US architect Wallace Neff had invented a similar pneumatic formwork system for creating low-cost domed housing; however his process involved covering the inflated fabric sphere with steel reinforcement and subsequently spraying it with concrete. Bini, by raising the weight of the concrete with air, had not only eliminated the need for costly and time-consuming formwork, but also made the construction process safer, with no need to work at height.

In the 1970s, Dante Bini went on to develop a smaller-scale version of his construction system, which he named Minishell. An 8 x 8m (26¼ x 26¼ ft) plan form was rationalized to a square base with mitred corners, which significantly simplified the reinforcement system and created openings at each corner. Bini produced multiple small buildings for use as holiday homes that were made with a very small amount of concrete – less than 5m³ (176 cubic ft) – to form the structural skin of a 64m² (698 sq ft) building. Bini has subsequently experimented with fibre-reinforced cement for a small shelter system, and developments in lower-carbon concrete and new reinforcement systems mean that Bini's invention from over 60 years ago remains relevant as a low-carbon alternative to more conventional construction. Dante's son Nicolò Bini continues to develop the technology by creating larger free-form plan layouts with inflatable formwork and low-cost emergency shelter accommodation prototypes.

BELOW Five-hour construction sequence diagram of a Binishell, produced by Dante Bini.

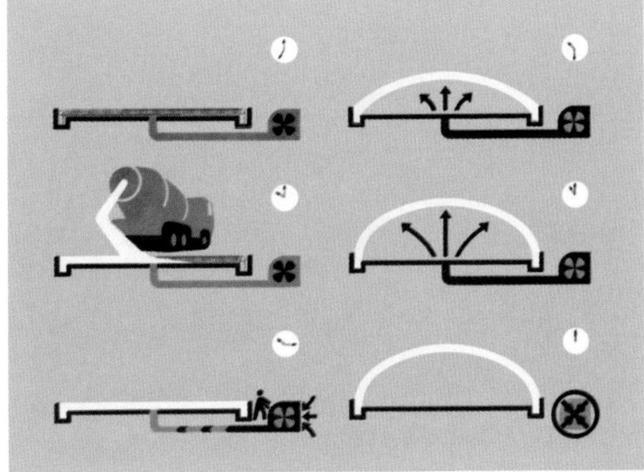

Poikilohydric (bioreceptive) living walls

Marcos Cruz, Richard Beckett, Sandra Manso, Brenda Parker, Chris Leung, Bill Watts (Bartlett School of Architecture, UCL)

KEY DATA

Compressive strength: Porous magnesium phosphate concrete (produced at UCL, tested at BRE) – 23.35 N/mm². Corkcrete (produced and tested at the University of Coimbra) – 3–5 N/mm²

Led by Professor Marcos Cruz at the Bartlett School of Architecture (UCL), this research project explores the potential of bioreceptive building surfaces that, in addition to being a wall as part of the climatic enclosure of a building, will also positively interact with their local environment and attract plants and support biodiversity. Whereas modern construction is predicated on creating strong and impervious surface treatments for long life and durability, Cruz asks whether these same building surfaces might also actively attract and store water, and the poiki-lohydric plants of algae, moss and lichens, without compromising structural and material integrity.

To improve the bioreceptivity of the cementitious building surface and encourage this plant growth, two broad avenues of research were pursued: firstly into the material substrate itself, with new mixtures of

'All the externally exposed surfaces of buildings and urban infrastructures, from blank walls and façades to roofs, retaining barriers and fences offer vast quantities of area to absorb and store water… Hydrophilic design allows us to take advantage of plants that will help us improve the storm-water management of façades and increase absorption of CO_2, nitrogen and pollutants while emitting significant levels of oxygen.'

Professor Marcos Cruz, Bartlett School of Architecture (UCL)

aggregates and binders to create a more hydrophilic (water attracting) surface; and secondly looking at the surface morphology, with different techniques of mould-making and casting used to create highly shaped and textured panels that can provide habitats (crevices, ridges and hollows) for the poikilohydric plants. Initially a test rig was designed and built on a site adjacent to London's Euston Station, where 18 large panels of different surface textures and materials were held above ground and studied for bioreceptivity, looking at surface temperature, moisture absorption and colonization rate for mosses. Three types of elaborately formed panel were tested, with three of each, using two types of material – Portland cement and magnesium phosphate cement.

Panels have subsequently been installed in St Anne's Catholic Primary School, East Putney Underground Station and Camley Street Natural Park in London, and it is interesting to see how they have variously performed in symbiosis with their local environment. It is also fascinating to see how the geometry of the panel surface has evolved, with later iterations having a more angular, branching system cut into the

surface. A later version of the precast cementitious panel has been developed with the University of Coimbra, Portugal, containing cork, which minimizes evaporative water loss from the panels.

This rigorous design research project might help in a re-imagining of 'green' or 'living' wall technology that does not require an irrigation system, while locally storing water and mitigating water run-off as well as remediating (cleaning) air and providing local climate control through evaporative cooling. If we study the vernacular palette of materials and construction technology anywhere in the world, it is clear how these materials and constructions interact with their local environment: living and breathing, rather than homogenous and inert.

OPPOSITE In situ testing of panels made with different textures and materials to study plant growth.
RIGHT Installing the living wall at St Anne's School in London, UK.

House of Cores: 3D-printed concrete

Leslie Lok and Sasa Zivkovic (HANNAH)
Location: Houston, Texas, US

KEY DATA

Embodied carbon: 360kg CO_2e (reduced from 406.6kg CO_2e)

Compressive strength: 34.47 N/mm²

Fire performance: Non-flammable and non-combustible

Fabrication process: The aggregate size used in the material formulation was increased from 4.76mm to 9.53mm – moving from 3D-printed mortar to 3D-printed concrete

For their much feted Ashen House experimental building prototype in Ithaca, New York (2019), designers and academics Leslie Lok and Sasa Zivkovic of HANNAH combined 3D-printed concrete and advanced digital fabrication and robotics to utilize previously unusable ash dieback timber. With their recently completed 'House of Cores' project, HANNAH again employs a hybrid approach to the structural and material construction of this innovative family home, with the eponymous cores made from thinly piped concrete that support timber floors and roof. The 'House of Cores' is a two-storey family home in Texas and the first multi-storey printed struc-ture in the US. The 'printed' concrete is delivered from a house-sized gantry that straddles the site, and the designers have sought to embrace the design potential of mass-customization that this fabrication process supports, with bespoke designs for storage, lighting and other internal features.

The key 'printed' concrete elements are the five structural cores that contain bathrooms, stairs, fireplace, storage and service spaces. The cores are fabricated from continuously printed walls built up from a special concrete mix, with a layered piping

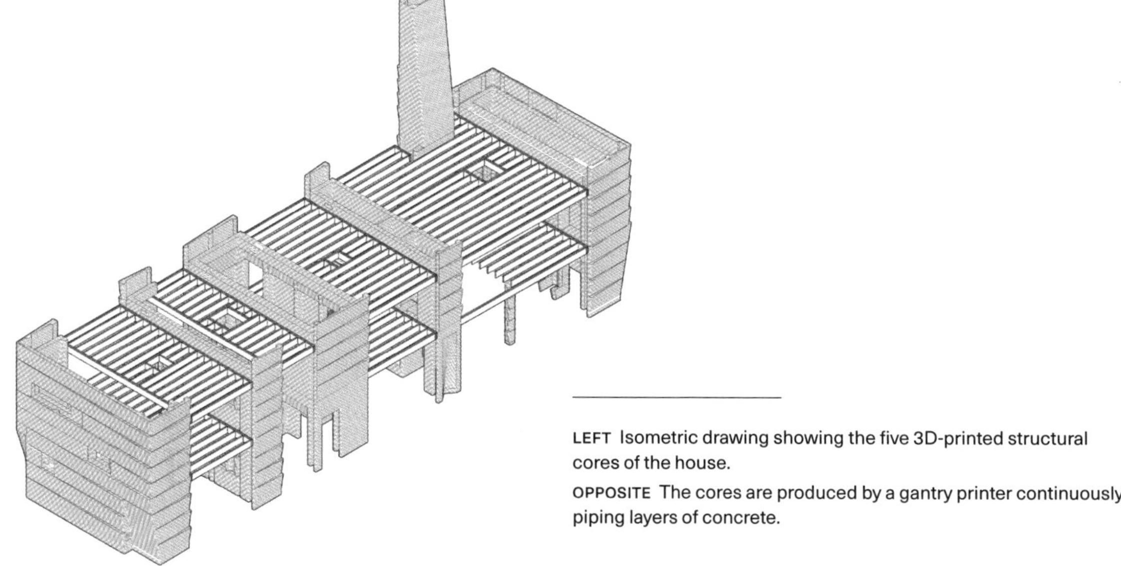

LEFT Isometric drawing showing the five 3D-printed structural cores of the house.

OPPOSITE The cores are produced by a gantry printer continuously piping layers of concrete.

'House of Cores' design strategies aim to increase the impact, applicability, sustainability and cost efficiency of 3D printing for future residential and multi-family buildings.'

Leslie Lok and Sasa Zivkovic, HANNAH

process more akin to cake icing than construction. These thin concrete surfaces mark out the outer surface of the cores, creating a hollow diaphragm wall cross-connected with steel reinforcement placed between layers. The wall voids also incorporate beams and embedded columns where required. These printed walls are tied to the building foundations with chemically anchored steel dowels. The printing process allows for lateral undulations in the walls, which the designers have used as corbel brackets for floor beam supports and for creating deeper window reveals.

The project is a collaboration with PERI, who manufacture formwork and scaffolding systems, and contractors CIVE, and the concrete is printed using the COBOD BOD2 construction printer. COBOD is a joint venture between PERI, General Electric, Holcim and Cemex (two of the world's largest concrete companies). The printer's simple modularity has enabled its use on this tight urban plot. The obvious environmental benefits of this printed (or additive manufacture) process are resource efficiency and the ability to create complex forms without the need for complex (and thus un-reusable) formwork during the construction process.

Prototypical projects like the House of Cores are invaluable as a testbed for new ideas about architecture and construction and demonstrate how we can usefully embrace technological change for improved human comfort, and economic and environmental sustainability. Thoughtful and inventive designers HANNAH continue to skilfully navigate a future design trajectory that employs new technology as part of a new architectural approach and not simply a technocratic exposition.

OPPOSITE The house was designed as a two-storey single-family home incorporating 371m² (4,000 sq ft) of living space.
RIGHT External walls are doubled up to prevent cold bridging, and internally filled with recycled foam for thermal insulation.

Zero-carbon concrete

Liv Andersson (Biozeroc)

KEY DATA

Embodied carbon: 0.28 kg CO_2e/kg

Compressive strength: 5–35 N/mm²

In the quest for a zero-carbon concrete, Biozeroc has successfully produced a 100% cement-free concrete. This new climate-tech start-up company has managed to literally grow cement-free concrete components with a method inspired by nature and the growth of coral. In a process called 'bacterial mineralization', bacteria is added to calcium and carbonates to create a natural glue that binds aggregates together, mimicking the cementitious properties of cement. This process of biomineralization is not dissimilar to the way in which invertebrate species such as oysters form shells and mammals produce bone, teeth, claws and nails.

Founded in 2021, Biozeroc was established with a goal to decarbonize the construction industry and 'provide access to carbon neutral construction materials, at scale, across the globe.' Biozeroc's process of bacterial mineralization operates at room temperature and is free from cement or any cement replacement materials such as GGBS or fly ash. Biozeroc's cement replacement material uses bacteria to precipitate the growth of calcite, and when founder Liv Andersson read about this technology being used in self-healing concrete, she wondered whether it could be scaled to bind aggregate at scale and produce concrete. Biozeroc's process is currently being patented, but Andersson explains that the chemicals or reagents used to precipitate this cementitious type process are widely available and can be sourced globally and at scale: 'If we succeed in sourcing feedstock chemicals from waste sources, the resulting concrete could be carbon-negative.'

The Biozeroc samples that have successfully been produced are smooth, dense and less porous than concrete, almost like a type of polished stone, and it is hard not to imagine this new substrate as a finishing or facing material.

LEFT Different types and colours of aggregate are bound together in a biomineralization process to form a type of zero-carbon concrete.

'In my old job, I worked to reduce the carbon embodied in new buildings and, sure, good design can significantly reduce a carbon footprint. But for me, it was not enough. We need new approaches, new materials, if we are to go further and get to a zero-carbon construction industry.'[9]

Liv Andersson, Biozeroc

Funded by the UK government's 'Innovate UK' scheme, Biozeroc has recently received substantial investment to establish a research partnership between the University of East London and industry partners including Ibstock (one of the UK's largest brick and block producers) and Aggregate Industries (part of the global cement and construction firm Holcim). This partnership certainly indicates how seriously the larger construction industry is taking the possibility of low-, zero- or even negative-carbon cement and concrete.

____

BELOW Inventor Liv Andersson and her Biozeroc team experiment with new substrate mixtures in the lab.

Essential Homes Research Project

Norman Foster Foundation and Holcim

The Essential Homes Research Project is
a collaboration between the Norman Foster
Foundation and construction materials multi-
national company Holcim. As part of the Venice
Architecture Biennale 2023, a full-size mock-up
was constructed at the Marinaressa Gardens,
Venice. The foundation identified that emergency
shelter and housing often have to last years, and
in some cases decades, and that while the tents
that are commonly deployed serve an important
immediate need, a more medium-term solution
is required.

'How can we ensure everyone,
including some of our world's most
vulnerable populations, can have
access to decent living conditions?'

Norman Foster

RIGHT   Rendering showing the layout of parts and the
construction sequence of the building.

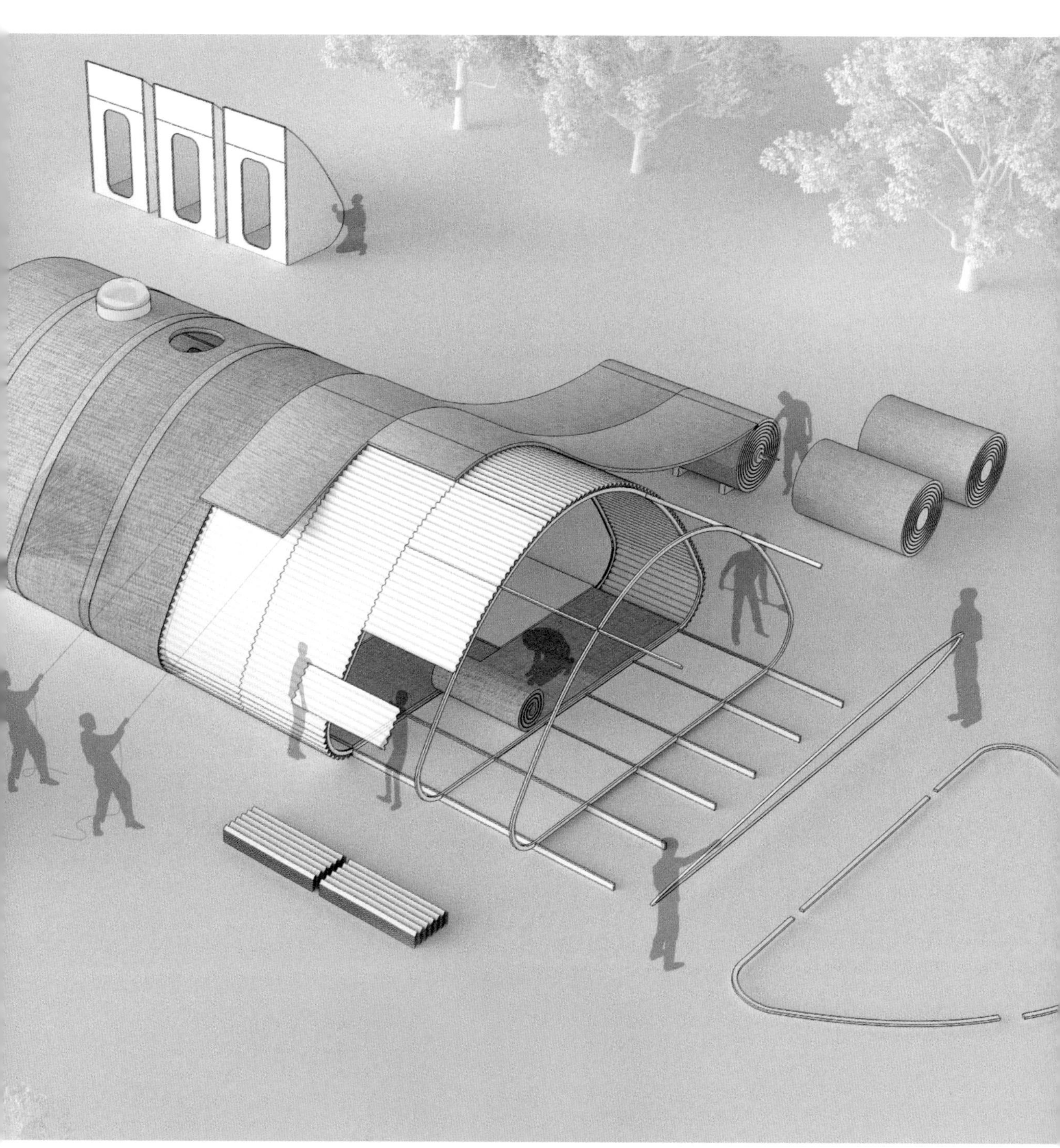

The building is an extruded tube 9m long x 6m wide and 3.4m high (29½ x 19½ x 11 ft); the structure is supported by a series of thin steel frames which define the cross-sectional shape, a permanent form-work of corrugated cement board, and sheathed by a protective covering of a product called CCX, which is designed and manufactured by Concrete Canvas. Concrete Canvas is a waterproof geotextile consisting of two layers of woven fabric with a cementitious core, and once wetted the cement is activated and will fully harden with 24 hours. The Concrete Canvas CCX skin is only 10mm (⅜ in) thick, and when bonded to the corrugated cement board and internal frames the total thickness of the external envelope is little over 100mm (4 in) thick. CCX is mostly used for lining irrigation channels and other water management projects, and as such its waterproofing credentials are proven and BBA certified. As a part of the research project Holcim also supplied a lower carbon cement for the CCX production, reducing embodied energy by 20%. While the high carbon emissions of cement are well known and discussed throughout this book, the invention of materials like Concrete Canvas vastly reduce cement use while harnessing some of its most important properties of structural strength and fire resistance.

Concrete Canvas was invented and developed by Peter Brewin and William Crawford while studying Industrial Design Engineering at Imperial College and the Royal College of Art in London. Their first project in 2004 was coincidentally an emergency shelter made from Concrete Canvas that was shaped by inflatable formwork. Concrete Canvas has since developed into a multi-million-pound business based in South Wales, UK.

OPPOSITE Full-sized structure on display at the Venice Architecture Biennale in 2023.
BELOW Kit of parts diagrams.

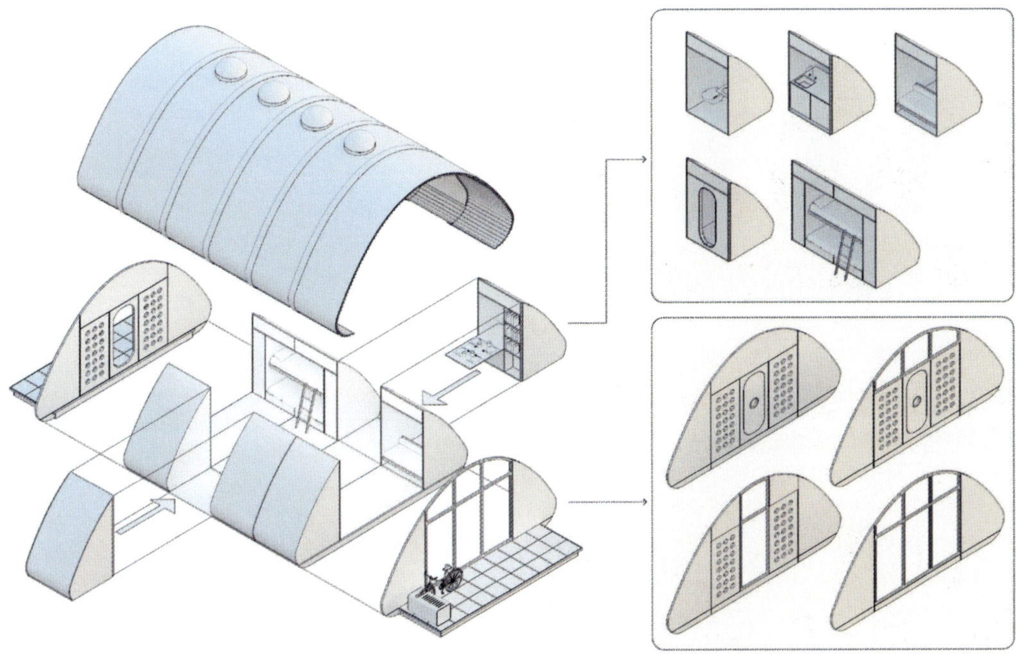

Trabeculated panels

Enrico Dini (D-Shape)

KEY DATA

Compressive strength: 10 N/mm² (The relatively low compressive strength is compensated by a steel exoskeleton for the panels products)

D-Shape was established in 2007 by inventor Enrico Dini. He had been working at Italian scooter company Piaggio in the early 2000s when he first saw the potential for 3D-printing technology, and speculated on whether it could be scaled for building-size components or indeed whole buildings. Dini registered the first D-Shape patents in 2005 and founded industry-pioneering Monolite UK in 2006 and D-Shape in 2007. In 2008 he built his first large-scale printer with his brother Riccardo, and produced his first, now iconic, large-scale printed structure – the Radiolaria Pavilion, which has been described as the world's first 3D-printed architectural structure. Designed by architect Andrea Morgante of London-based Shiro Studio, the pavilion is made from ground marble dust bonded together with a binder that is sprayed via a huge XYZ gantry system. The additive 'printing' process works by building up layers of powder and then solidifying (printing) areas with the binder. When the process is complete, the surrounding loose (unbonded) powder is simply brushed away to reveal the solid structure. The loose powder also acts to support the structure throughout the print/ fabrication process.

Dini has continued to work with Morgante, and they have subsequently developed a series of three types of trabeculated panels intended for intertidal application. The panels are designed like a kind of artificial reef to boost growth of bivalves such as oysters and clams in harbours, enhancing marine biodiversity and cleaning water through the natural filtration capabilities of bivalves. With the environmental collapse of some coral reef systems, there is great demand for artificial reef systems and the biodiversity that they provide. Other artificial reef systems do exist and are typically made from precast concrete, but it is hard to create the geometric

LEFT 3D-printed eco-concrete trabeculated panel prior to its installation in Genoa's harbour as part of an artificial reef for marine life.

complexity and large (and textured) surface area with standard concrete prefabrication processes. With Dini's process, complex 3D forms are achievable, as are finely textured bio-receptive surfaces on which organisms can grow and thrive.

The panels are made from a specially formulated eco-concrete with a very low pH level, using quarried aggregates and pozzolanic and magnesium-based cementitious material. In Dini's latest artificial reef project for ENEA (Italy's national agency for new technologies), quarried aggregates will be replaced with crushed waste shells (calcium carbonate) and used as aggregates (70% by volume). Dini continues to develop the geometry of the reef panels, and by

increasing surface area and using waste seashells from other aquaculture processes improves the circular economy and biocompatibility (with marine life) of the material. D-Shape have recently printed 60 large-scale panels, which are currently being installed in Genoa harbour. Dini's dream of 3D-printing houses has been realized, albeit as habitats for bivalves.

———————————

BELOW The 3D 'additive' printer applying layers of sprayed bonder to locally solidify a powdered cementitious material. When the process is complete, excess powder is removed leaving the printed 3D form.

2.3 Stone

'[The industry] routinely takes limestone at 100–200 N/mm^2 strength and crushes it up, gets a load of gas and burns it, turns it into cement, gets freshwater which is in short supply and sharp sand which is in short supply, puts up formwork and falsework, adds reinforcement, back-propping, vibrates it, puts releasing agent on it, strikes the back-propping and gets the formwork off and, hey presto: a material (reinforced concrete) of 40 N/mm^2!'

Steve Webb, Webb Yates Engineers

Post-tensioned stone

Webb Yates Engineers

KEY DATA

Embodied carbon of 'general' stone:
0.079kg CO_2e/kg

Compressive strength of limestone:
100–200 N/mm²

Structural engineer Steve Webb has led a re-evaluation or rediscovery of stone as a contemporary structural material. Working with architects and fabricators including Amin Taha (Groupwork) and The Stonemasonry Company, he has shown how stone can be reintroduced into the designer's palette and understood as a low-carbon and locally sourced construction material. In 1992 structural engineer Peter Rice, working with architects Martorell Bohigas Mackay (MBM), created the Pavilion of the Future for the Seville Expo. This audacious engineering set piece saw the construction of a wall of eleven 37m (115 ft) high arches made with a delicate filigree of precision-cut pink granite components that were held in place by an exoskeleton of tensile steel rods. Rice had brilliantly demonstrated how the compressive load (the literal weight) of the structure is easily absorbed through the stone, with the stability of the structure provided by the external tensile steel grid.

BELOW Post-tensioned stone beam fabricated from stone elements held together (in tension) by internal steel cables, part of 'The New Stone Age' exhibition, 2020, curated by Groupwork, The Stonemasonry Company and Webb Yates.

BELOW The Formby Stair, a 22-tread two-storey stone staircase with interlocking elements threaded with post-tensioned cables.

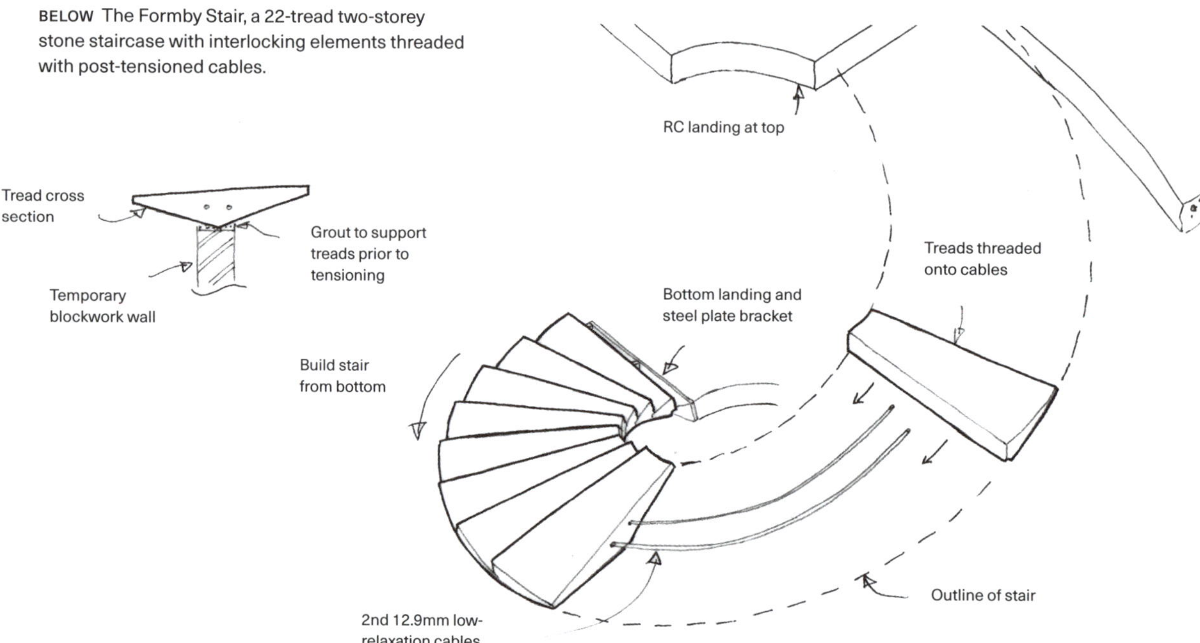

Much of Webb's work with stone utilizes Rice's technology, but instead he embeds the tensile elements (steel cables) within the stone. In the same way that steel bars provide reinforcement in concrete, by inserting steel cables within pre-drilled holes in stone, once post-tensioned they lock in the tensile forces. This simple process allows the designer to transform a low-tensile material like stone – good for bearing weight, but bad for spanning – into a material with better structural properties than reinforced concrete. With Webb's system it is now possible to create a thin-plate stone floor from pre-cut stone components and the simple addition of wire rope that is then post-tensioned.

In a recent talk, Webb came prepared for questions about the relative sustainability, availability and cost of stone. In many countries it is a local material with different types and strengths available across the country. Mining (or quarrying) stone is environmentally low-impact and waste can be used as aggregate and other construction products. The tensile component is steel cable (rope) which is light and only requires simple tools and hydraulic jacks for post-tensioning.

Key projects include 15 Clerkenwell with Amin Taha, where a mixed-use six-storey building is supported by solid limestone load-bearing columns that support reinforced concrete floor plates. Each column is designed for specific calculated loadings and sized accordingly. To save cost and workmanship, the finish of the stone is largely as it was when it was cut from the quarry. Split naturally along the bedding planes of the stone, this provides a visual contrast between natural rough edges and worked smooth edges. Other projects include a series of breathtaking helical stone staircases like the Formby

Stair, a 22-tread two-storey stair made from individual interlocking stone components threaded together and held in place by twin embedded and post-tensioned cables. For the 2022 Royal Academy of Arts Summer Exhibition, Webb Yates, working with The Stonemasonry Company, created 'Equanimity', an 11m (36 ft) long post-tensioned stone beam made from eleven 1m (3¼ ft) long components of Portland stone and upcycled granite. The stone had been rejected by the quarry for visual blemishes, and the granite was salvaged from a London dockland building's demolition and recut. Webb had wanted to create a visual demonstration of how structural stone can be used in place of concrete, saving huge amounts of carbon in the process.

RIGHT The components of the 'Equanimity' stone beam were held together with three post-tensioned steel cables and held aloft by a Douglas fir frame.

2.4 Metals

'Early studies performed by Symmetrys suggest that reuse of steel can cut embodied carbon by 80% in comparison to typical procurement routes. Although reusing steel reduces embodied carbon more than recycling it, the demolition industry currently follows a recycling first approach, which makes finding suitable sections from reclaimed material for reuse challenging due to the limited marketplace.'

Symmetrys, Structural and Civil Engineers

Maisons Tropicales

Henri and Jean Prouvé
Location: Brazzaville, Republic of Congo
(later in France, UK and US)

One of Jean Prouvé's most important technical innovations was as an early adopter of the hydraulic brake press, which could fold, form, punch and perforate sheet metals (typically steel or aluminium). The folding and forming of thin gauge metals creates great structural strength with lightweight materials, and Prouvé's construction and fabrication methods appear more closely aligned with the aeronautical and automotive industries. The structural frames in a Prouvé building are not fabricated from heavyweight hot-rolled steel sections, but composed of hollow folded sheet metal profiles and forms which optimize structural strength and lightness through the monocoque or stressed skin principle.

Prouvé originally apprenticed as a blacksmith and opened his first studio at the beginning of the 1920s making locks and ironmongery. Arguably the most productive period of his working life was when he had his own fabrication facility at Maxéville, near Nancy, which he opened in 1947. Maxéville contained a design office and a large factory space where some of Prouvé's most celebrated works were produced. He is as well known for his furniture designs as his architecture, and the facility at Maxéville allowed him to fabricate furniture, doors, windows, cladding panels and whole building structures. Prouvé often worked with other architects, and his fabrications can be seen in the work of Le Corbusier, Oscar Niemeyer and Candilis, Josic and Woods.

The Maisons Tropicales, or Tropical Houses, were commissioned as a pair of buildings by Studal, an engineering firm specializing in the use of lightweight alloys. These were designed by Jean Prouvé and his brother Henri, fabricated at Maxéville in 1950 and shipped to Brazzaville in the Congo (now the Republic of Congo). The plan dimensions of the buildings, designed as offices for Studal, were 10 x 20m (33 x 66 ft) and 10 x 14m (33 x 46 ft), and the 140m² (1507 sq ft) building weighed less than 10 tonnes (11 US tons). The buildings were prefabricated as a series of components small enough to fit

---

BELOW Reassembly of Prouvé's Maxéville design office, displaying the 'tuning fork'-shaped supports made from folded sheet metal that he would later use in the Maisons Tropicales.

'Architect? Engineer?
Why bring up this question?
Isn't our job simply to build?'

Jean Prouvé

into an aeroplane and light enough for two people to handle and assemble by bolting together. The success of the demountable qualities of Prouvé's buildings is evidenced by the relative ease of their relocation, and in 2001 one of the Tropical House buildings was brought back to Paris, restored and reassembled. It has subsequently been exhibited in London and the US, where it was sold for six million dollars. The Tropical Houses are designed on a 1m (3¼ ft) grid system and consist of a series of ground beams connected to Prouvé's recognizable portal frame system of 'tuning fork'-shaped supports fabricated from folded sheet steel. Most of the other structural components are made from aluminium sheet, with no element longer than 4m (13 ft), which corresponds to the capacity of the brake press, nor heavier than 100kg (220 lbs), for ease of handling.

Alongside the environmental benefits of incredible material efficiency, the Tropical House is also an exemplary model of how to mitigate the extremes of the hot and humid tropical climate. The outer light-reflecting building skin consists of adjustable brise-soleils, which shield the office accommodation from direct sunlight and are separated from the inner insulated skin of sliding doors (with rotating vents) and fixed panels. The profile of the roof wings is designed to exploit the stack effect, with the whole structure held above ground on small pilotis to stimulate good air flow and ventilation.

RIGHT Assembling the Tropical House from the prefabricated kit at Maxéville, Prouve's construction facility in France, in 1951.

Trumpf Footbridge: lightweight shell construction

**Schlaich Bergermann Partner
and Barkow Leibinger
Location: Ditzingen, Germany**

KEY DATA

Dimensions: Length 28m (92 ft), width 10m (33 ft), width of walkway 2.2m (7¼ ft), height of railing 1.1m (3½ ft)

Thickness of steel shell: 20mm (¾ in)

Weight: 21 tonnes (23 US tons)

Embodied carbon: 2.08–6.82 CO_2e/kg (depending on recycled content from 30–85%)

Compressive strength: 600–700 N/mm²

While this highly crafted and exquisite piece of precision-engineered steel might not appear to be a model of sustainability, it is in fact an object lesson in material efficiency and in doing more with less. Fabricated from corrosion-resistant duplex stainless steel in six sections, the bridge was welded together adjacent to its final location and craned into place in one piece.

Mike Schlaich, like his father, the renowned structural engineer, Jörg Schlaich, uses form-finding to help model and develop structural form. In an early improvised model, a tensile net bag (the type you buy oranges in) was stretched over an arched timber sheet that spans between two embankments. The design team combined these forms and geometry to define a single doubly curved surface. The double curvature gives incredible strength, creating a shell-like monocoque surface, which engineer Heino Engel described as 'form active' in his book *Structure Systems* (1967). A sheet material of only 20mm (¾ in) thick spanning almost 30m (100 ft) seems like an impossibility, but by learning from the form-finding model experiments and careful three-dimensional shaping, a bridge is formed.

BELOW Elevation drawing of the steel bridge structure showing the impressive span-to-depth ratio and how slender the deck is.

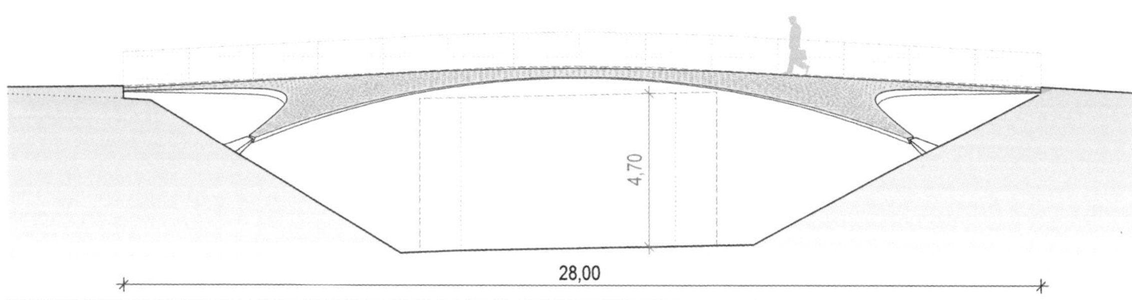

Bending the perforated steel sheet in a shipbuilding yard.

Attaching the spherical 'ball joints' that support each of the four legs. This type of structural joint allows for the thermal expansion of the steel bridge.

The bridge is lifted into place in a single piece. Note the steel leg and ball joint connection to the embedded concrete abutment.

'Laser-cut holes in the shell not only make the structure lighter and translucent, but also the remaining steel work visualizes the path of the principal membrane forces, thus making the bridge more understandable.'

Mike Schlaich, Schlaich Bergermann Partner

Integral to the realization of this project was the use of boat builders for the forming and shaping of the sheet steel. Nautical architecture is by nature curvilinear, and the forming of complex doubly curved surfaces is well understood. Using pneumatic presses, steel sheet is deformed at a series of points across a surface, pushing the sheet steel into the desired shape. This process is very similar to that of 'wheeling' steel using the 'English wheel'

for the fabrication of handmade car body panels. The fabrication of the bridge began in the Netherlands with 6 x 3m (19½ x 9¾ ft) sheets of stainless steel, laser-cut to shape and perforated with a pattern of cut-outs defined by finite element (FE) analysis. The FE analysis determined which areas of steel could be removed without compromising the structure, and acts like a visual stress map showing the flow of forces. This removal of material also minimizes the self-weight of the structure, with less of its own weight to carry. Across the walkable deck the stress mapping of the bridge continues, with a pattern of over 13,000 holes drilled in the deck as well as a blasted anti-slip finish.

The edges of the single sheet structure are folded down to create a 200mm (8 in) deep collar which helps to stiffen the free edge. These collars meet at the four main support points. The bridge abutments at each side vertically support the bridge edge but allow for horizontal expansion and contraction. The balustrade is held in an aluminium channel, bolted to a vertical steel plate and welded to the surface of the bridge.

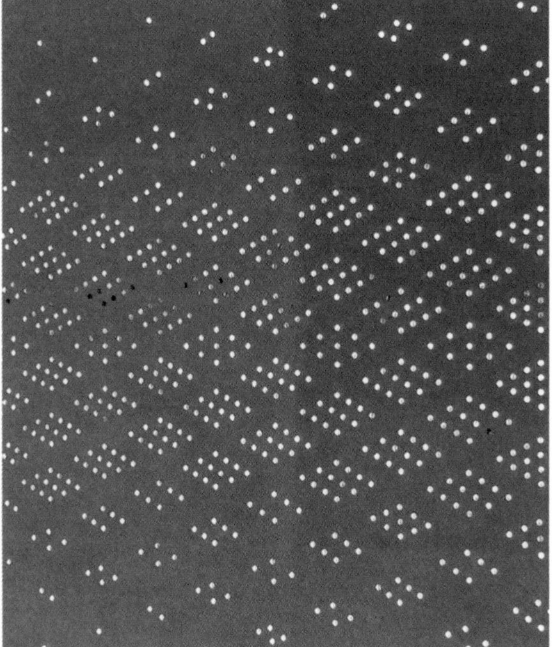

LEFT The holes were filled with clear glass plugs, which act like lenses or simple fibre optics to create an illuminated soffit.

OPPOSITE The balustrade is created in a white low-reflection glass for minimal visual impact; a stainless steel sphere connects the steel bridge to concrete foundations.

Study Pavilion TU Braunschweig

Gustav Düsing and Max Hacke
Location: Technical University of
Braunschweig, Lower Saxony, Germany

This study pavilion at TU Braunschweig in Germany presents a creative use of small hollow steel sections to create a two-storey study centre. Visually lightweight and without any identifiable cross bracing, this 1000m² (10,764 sq ft) building is conceived as a kit of parts for ease of assembly, disassembly, and future relocation to another site. The illustrated 'Material System' shows the fabricated connection node onto which 100mm (4 in) square hollow sections (SHS) are slotted and bolted. Steel angle sections attached to the horizontal members support the cross-laminated timber (CLT) floor panels, which are simply dropped into the frames. A 3 x 3m (9¾ x 9¾ ft) grid is maintained throughout, creating an open-plan learning environment but serviced by the SHS frames that also act as conduits for electrical

BELOW Diagram shows the kit of parts that make up the building, including fabricated connection node, hollow sections, window frames and CLT floor panels.

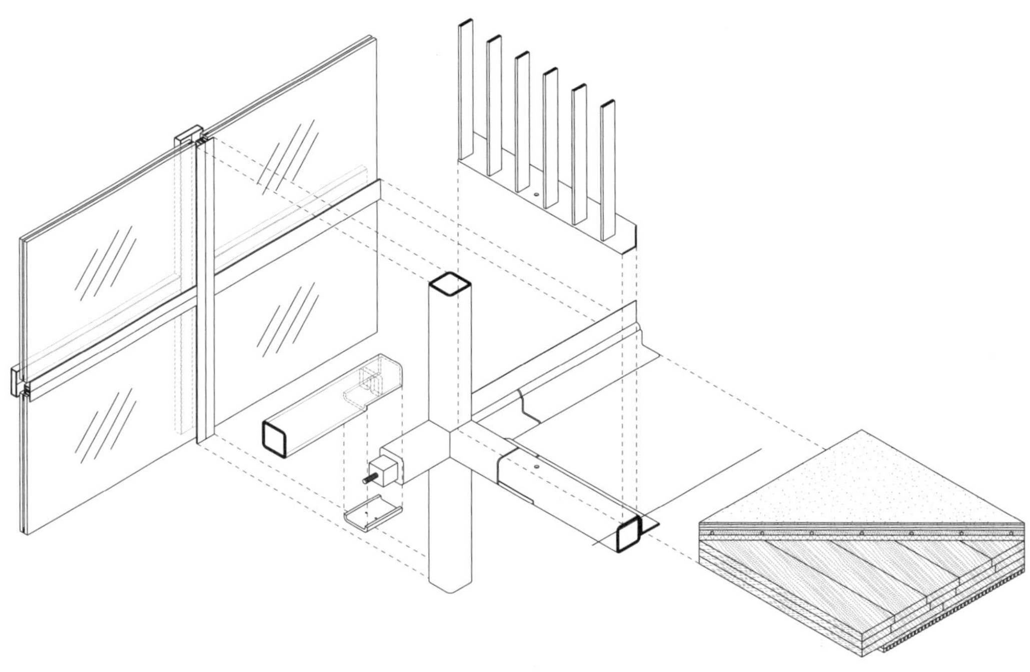

'The student pavilion is an example of sustainable and flexible campus architecture. The steel-wood hybrid construction is completely demountable and allows for easy assembly and disassembly.'

Gustav Düsing and Max Hacke

and AV cabling. A 3 x 6m (9¾ x 19½ ft) pod on the ground floor is the only enclosed space across the two floors, with additional visual and acoustic separation provided by bright yellow acoustic curtains.

There is no mechanical ventilation, and the building is naturally vented with openable window vents and a large openable circular rooflight. The building is heated with underfloor heat provided from the local district (community) heating, and the slab cooled in the summer with a ground source heat pump. The entire square plan building is galleried to the first floor, which provides good solar shading to prevent overheating in the summer while allowing solar gain from low winter sun. The fully glazed exterior is modular and screw-fixed to the steel SHS uprights, and easily demounted or replaced. The roof is formed by a profiled steel deck and a thick layer of foamed glass insulation. The seemingly 'invisible' cross-bracing is provided by the nine steel staircases, which give the lateral stiffness and bracing for the structural framework.

As part of this competition-winning design, the architects created a visual parts list to itemize all the materials required for construction of the project, and thus minimize waste and accurately quantify materials and components. While it is important to understand that the production of steel is currently a high-carbon activity, it is also important to note that steel is 100% recyclable and can also be easily reused by cutting, rewelding, etc. The steel industry is in the process of decarbonizing production, with a move to electric arc furnaces that re-smelt old steel into new as opposed to the primary production of steel from iron ore. Well-designed projects like this make great use of slim steel sections and showcase the versatility of this strong material with clever jointing systems and precision fabrication.

BELOW A glazed façade wraps the building, flooding the interior with light and connecting it with the outdoors. The roof overhang protects against unwanted solar gain in summer.

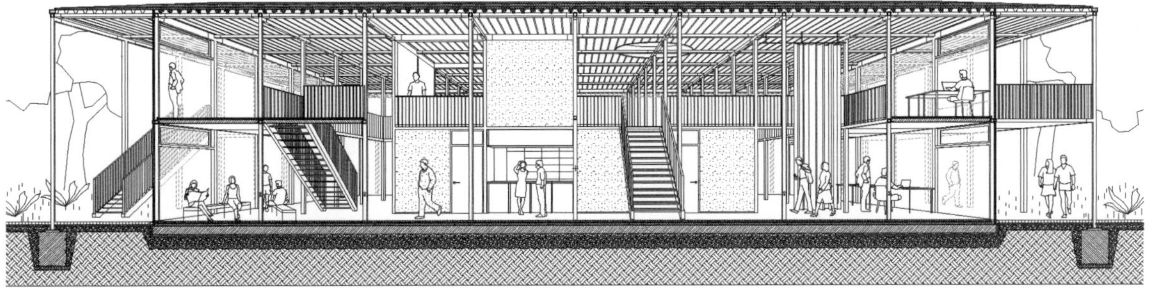

Thermobimetal

Doris Sung (DOSU Studio)

Architect and educator professor Doris Sung established DOSU Studio alongside her teaching and research role at the University of Southern California (USC). The studio explores passive and low-energy responsive façade systems that employ thermobimetals as the key dynamic (or moving) element. Thermobimetals are a laminate of two different types of metal with different coefficients of expansion – meaning that one metal expands more than the other with the application of heat. The metals are molecularly bonded through pressure so they cannot delaminate. Thermobimetals were used for decades for the thermal control of internal environments, albeit in compact format as the analogue thermostat switch operating domestic heating.

Doris Sung has taken thermobimetals 'out of the box' by creating a series of large- and small-scale installations that demonstrate how a thin perforated metal surface can change shape with the heat of sunlight. These responsive surfaces can help solar-shade or ventilate a space with a fish-scale-like structure that opens and closes in response to local heat. The advantage of using such technology is

BELOW Self-shading windows, with thermobimetal 'petals' within a double-glazed unit. Each petal is autonomous and reacts to local heat, mirroring the heliotropic behaviour of plant petals.

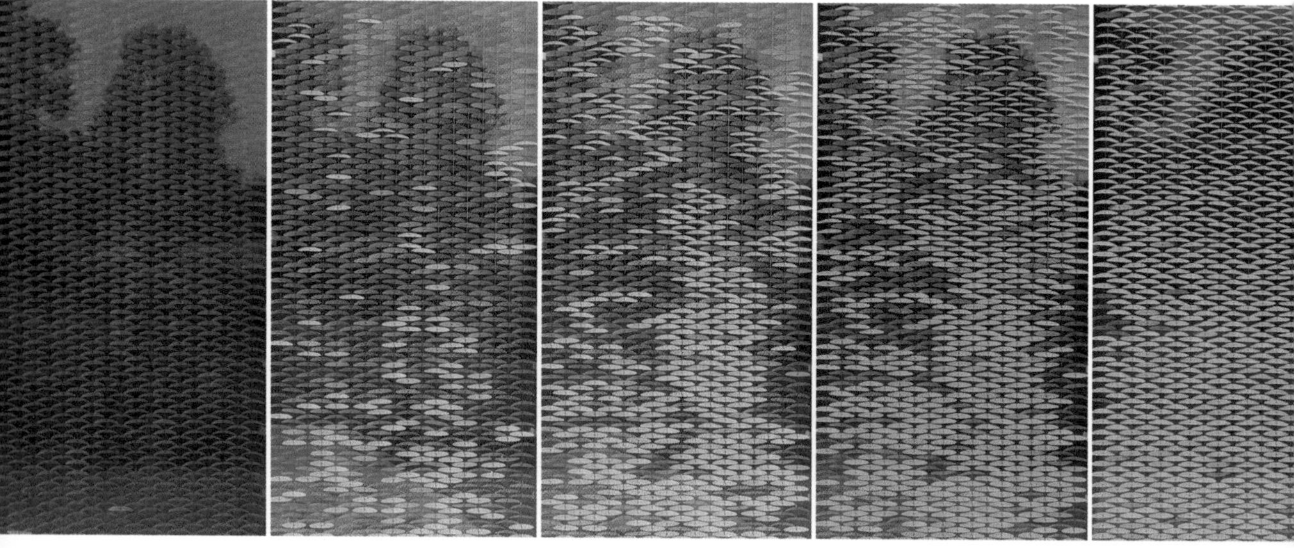

'What the market has defined as "smart" is just the gadgets … as architects why aren't we making smart walls, ceilings and floors?'

Doris Sung, DOSU Studio

---

that it does not require any external power supply or sensing systems, with solid-state systems changing with weather and local environmental conditions as required. Together with students, Sung has created sculptural installations to test and demonstrate both the beauty and operational effectiveness of the technology. In the Bloom project, sheets of bimetal are pattern cut and joined to create a suspended doubly curved monocoque form. The sheets are CNC-machined with scale-like profiles cut but still attached to the main sheet, and so mechanical connections are kept to a minimum with the integrity of the curved surface used to support and hold the moving parts. Bloom was the first time Sung and her students created an external structure made from bimetal, and it successfully demonstrated how the changing sun path would lift and close the scales of the structure, ventilating and shading as appropriate.

Sung has subsequently explored how thermobimetal technology might be integrated into more conventional window and roof components, and created a separate start-up TBM to develop InVert™ self-shading windows. Sung had observed that while low-e (low-emissivity glass) coatings were effective at mitigating solar heat gain, they are not great for human comfort and wellness as the world is tempered through 'dark glasses', as are the beneficial health effects of broad-spectrum daylight. InVert™ traps a filigree network of thermo-bimetal petals hooked on thin metal fibres within a double-glazed window unit. When the petals heat up, they flip to face and block out the sun; InVert™ is described as a 'biophilic, self-shading system' to reduce air-conditioning demands and energy in

use, thus reducing carbon emissions. This is an elegant example of usefully employing the mechanical properties of a given material as part of a responsive system for both visual delight and climate control within architecture.

BELOW Oculus is a thermobimetal aperture that opens and closes according to heat. Sung's team had the idea to develop the concept into a valve system for a wall or ceiling.

2.5 Glass

'Glass is an inert material that has the potential to be recycled in a closed-loop system indefinitely. Coupled with the short service life of insulating glass units (double and triple glazing) there appears to be a disparity between the material's potential and its current utilization.'

Graeme DeBrincat and Eva Babic (Arup)

The Burrell Renaissance Project: recycling construction glass

John McAslan & Partners with Arup
Location: The Burrell Collection, Glasgow, UK

KEY DATA

Embodied carbon: Recycling one tonne of window glass saves up to 300kg CO_2e/kg

Glass is a 100% recyclable material, but in the UK it is not being consistently recycled, with much of the UK's annual 200,000 tonnes (220,500 US tons) of post-consumer glass waste ending up in landfill sites or downcycled as aggregate. Construction waste accounts for an estimated 25–30% of waste in the EU and includes concrete, brick, glass, metals and plastics. A recent Deloitte study on sustainability identified that glass used in construction is rarely recycled into new glass products; however most flat glass produced using the float glass process could be recycled using existing glass furnaces.

BELOW If broken float glass is stored separately and not mixed with bottle and container glass, it can be successfully recycled as new architectural glass without any loss in quality.

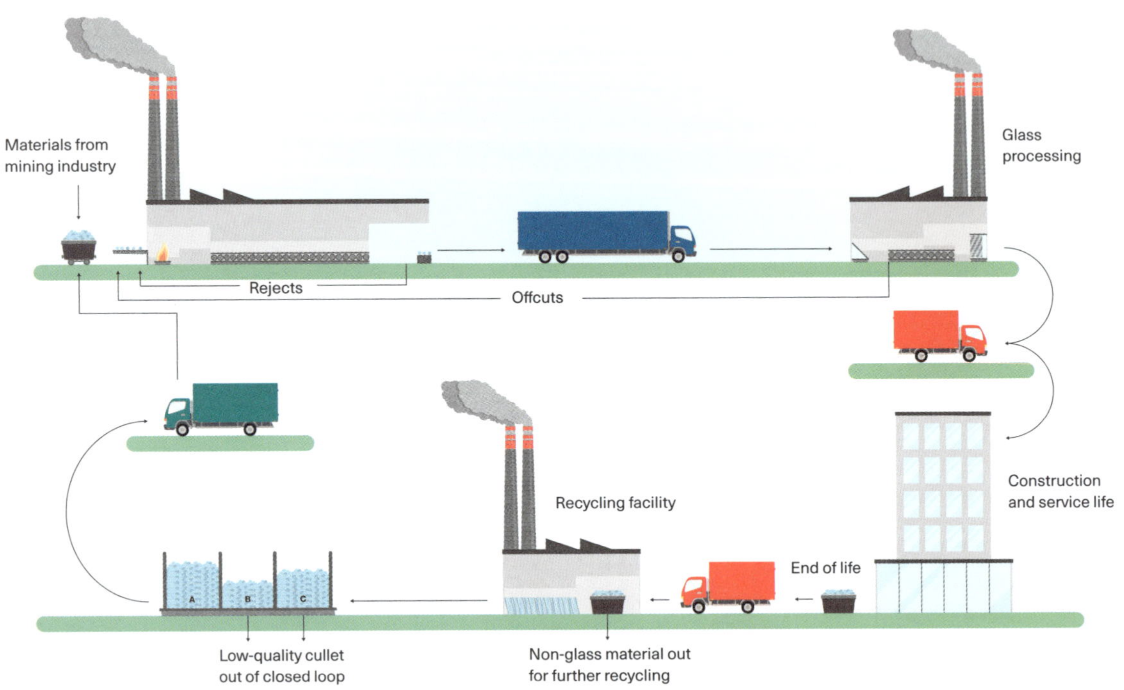

Materials from mining industry
Glass processing
Rejects
Offcuts
Construction and service life
Recycling facility
End of life
A
B
C
Low-quality cullet out of closed loop
Non-glass material out for further recycling

Nationwide Platforms

'The manufacture of one square metre of low-e double glazing
leads to the emission of 25kg of CO_2 according to NSG Pilkington.'

Graeme DeBrincat and Eva Babic, Arup

One of the key factors preventing this recycling is the contamination of waste glass streams; once glass has been broken up it is difficult to distinguish the different grades and quality, so if construction glass is recycled at all it is for lower-value products such as bottles and jars or aggregate (see Gent Waste Brick p.56). In their recent report entitled 'Re-thinking the life-cycle of architectural glass',[10] Graeme DeBrincat and Eva Babic from design and construction company Arup explain that a consistent recycling of waste building glass across Europe could save upwards of 900,000 tonnes (1 million US tons) of landfill waste every year, and simultaneously reduce demand for raw materials – the chief component (over 70%) of which is high-quality silica sand. Only the highest grade of silica sand can be used in the production of glass, and while there are still UK supplies of this mined material, they are finite and so other sources of imported silica sand will have to be identified. Globally, silica sand and gravel are the most extracted solid materials and the United Nations Environment Programme identified in 2014 that 'they are now being extracted at a rate far greater than their renewal.'[11]

Glass cullet is broken glass that is returned to the furnace for remelting. Cullet is already used as one of the ingredients of new float glass and is usually sourced waste from factory offcuts or local glass processors. EU glass manufacturers currently include 20–25% of glass cullet in the production of glass, which could easily be increased by up to 50% if correctly sourced and graded. Most flat glass products can be recycled, except for laminated glass that has a PVB or resin interlayer, although the automotive industry is currently developing delaminating technology so that this will be possible in the future. Again, it is other industries working within more stringent regulatory environments that are innovating in the improved reuse and recycling of materials.

In a recent high-profile building refurbishment project in Scotland, the restoration of Glasgow's Burrell Collection involved the replacement of much of the original glazing from 1981 to improve energy efficiency and building performance. Over 3120m² (33,583 sq ft) of double-glazed units were replaced and, working with glass suppliers, the glass was carefully removed from aluminium and steel framing, which was also reused and refurbished. The glass was recycled as glass cullet for new flat glass production and other glass products, which diverted 110 tonnes (121 US tons) of glass from landfill disposal, with an estimated saving from this process of 27.53 tonnes (30 US tons) of carbon dioxide. This landmark cultural project showed how any large-scale or planned refurbishment of buildings, the upgrade of public housing for instance, could see this process legislated for and repeated.

ABOVE A glass production circular economy 'closed loop' process diagram produced by ARUP.

LEFT After the frames are removed for reuse, the glass is separated for use as cullet in the production of new float glass.

Current Window: photovoltaic solar energy

Marjan van Aubel

Marjan van Aubel describes herself as a solar designer working between design, sustainability and technology. She became interested in solar energy, but questioned why solar panels are so ugly, or at least not successfully integrated into their physical environment. She has developed solar cells that use the properties of differing light frequencies (colours) to produce electricity. Van Aubel was inspired by reading *The Solar Revolution* by Steve McKevitt and Tony Ryan, where they explain that within one hour we receive enough sunlight to provide the world with enough electricity for an entire year.

Working with a technology called dye-sensitive (photovoltaic) solar cells – which employs a similar principle to that of the photosynthesis of plants, where the green chlorophyll converts light into sugar (or energy) for plants – van Aubel's coloured solar

BELOW Van Aubel's 'Current Window' features panes of glass containing dye-sensitized solar cells, converting energy into electrical current.

'I believe in solar democracy: solar energy for everyone, everywhere. My aim is to make all surfaces productive. I want to build houses where all the windows, curtains, walls and even floors are harvesting electricity… it's time for the transition from solar technology to solar design, where we use the power of beauty to help facilitate change.'

Marjan van Aubel

cells turn light into electricity. Different colours have different efficiency depending on their position in the colour spectrum, with red (at one end of the visible spectrum) being more efficient than blue (at the other end) in converting light to electrical energy. Van Aubel has produced products such as 'Current Table', which uses coloured solar cells as a tabletop, with batteries in the legs to store electricity, for charging small devices such as phones. More interesting architecturally is her 'Current Window' project, where she replaced the windows of a gallery in London's Soho with a contemporary version of stained glass, where visitors could come and charge their phones. Sunlight is collected by orange, blue and pink-dyed panes of glass containing dye-sensitized solar cells. The light-absorbing dyes cover particles of titanium dioxide that convert the energy into electrical current.

Van Aubel wants to make solar energy available to everyone and not just those who can afford a sustainable lifestyle. She also wants to utilize the power of good design to help alter the perception of this highly technical but thus far under-designed substrate so that it becomes part of architects' and designers' palettes. As ecologically responsive designers, we need to think about how we harness all aspects of the sun using photovoltaic cells and passive solar thermal heating (with thoughtful building orientation), and about the health benefits that good-quality natural lighting provides.

ABOVE Solar energy collected from the windows can be used to charge electrical devices such as mobile phones.

Re³: recycled glass bricks

**Faidra Oikonomopoulou and Telesilla Bristogianni (TU Delft); the Glass & Transparency Research Group
Location: Delft, the Netherlands**

KEY DATA

Flexural strength of recycled cast glass: 30–75 N/mm²

BELOW Diagram showing RE³ glass strategy.
OPPOSITE Overview of fractured cast glass samples; bending test of a glass beam made from recycled oven door glass.

The Re³ glass project name embodies the hierarchical waste cycle of reduce, reuse and recycle. Faidra Oikonomopoulou and Telesilla Bristogianni are researchers at Delft University of Technology working with the Glass & Transparency Research Group. In studying waste glass streams, it became obvious that despite the recyclability of glass, a great majority of glass waste is not recycled at all and ends up as landfill. While the packaging industry does a reasonable job in the reuse and/or recycling of food and beverage containers such as bottles and jars, other industries, such as construction, automotive and consumer electronics (phone, TV and computer screens) do not readily recycle their glass products. Oikonomopoulou and Bristogianni identified one of the key mitigating factors in the failure to recycle high-value glass to be the demanding tolerance factors for transparency and structural strength (demanded for float glass), which are compromised by impurities during recycling.

The Re³ project demonstrates how this waste glass can be easily transformed into a high-strength modular and reusable construction system.

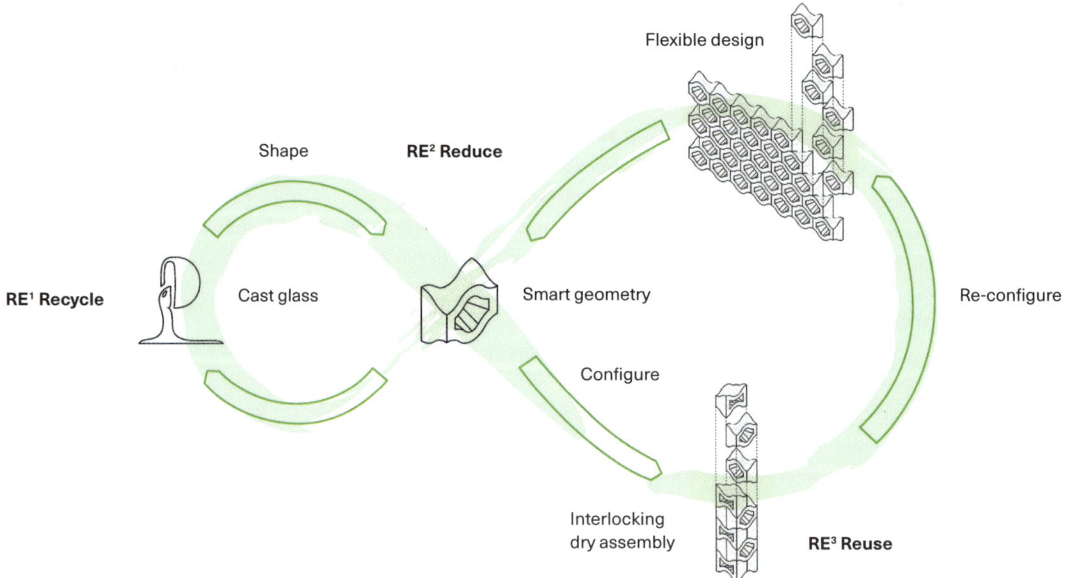

'Despite the common notion that glass is 100% recyclable, the majority of everyday discarded glass objects – apart from beverage and food containers – are neither reused nor recycled.'

Faidra Oikonomopoulou and Telesilla Bristogianni, TU Delft

Using both kiln-casting and hot-pouring techniques, waste construction, automotive and consumer electronics glass is transformed into interlocking glass block components of different geometries, transparencies, colours and textures. The fabrication process sees waste glass heated until molten and then poured into steel moulds of virtually any shape, and certainly not limited to rectilinear bricks, with the modular nature of the Re³ designs allowing for ease of assembly and disassembly. Glass has excellent compressive strength, and in creating larger cross-sectional components we can reimagine glass as a load-bearing and free-standing structure. As opposed to the float glass used for windows, we can also imagine a new kind of thick glass translucent wall which provides privacy, diffuse light and colour.

The TU Delft Glass & Transparency Research Group developed the technology for MVRDV's remarkable Crystal Houses shop façade in Amsterdam; informed by its traditionally constructed clay brickwork neighbours, a transparent wall of cast glass bricks is UV bonded by an invisible adhesive. While this project was commissioned by a high fashion retailer and the glass was not recycled, the project demonstrates the strength and robustness of glass brick construction.

BELOW Recycled glass panel; interlocking system of recycled window panes, crystalware, CRT screens and optical lenses.
OPPOSITE Two-component interlocking system of recycled low-iron window panes and artware; interlocking system of recycled glass artware, slowly cooled down to form glass ceramics.

Water-filled Glass (WFG)

Dr Matyas Gutai

Invented by architect and entrepreneur Dr Matyas Gutai, Water-filled Glass (WFG) is a technology that sees water introduced into the cavity of a double-glazed glass window unit. Designed to lower heating and cooling demands, this patented invention also provides local temperature control, preventing overheating in summer and heat loss in winter. Water is not a thermally insulative material, but it acts as a very efficient thermal store for heat and coolth.

Window glazing is the largest single source of heat gain and heat loss in a building. A window's thermal performance can be improved with the addition of layers and airgaps, with triple glazing becoming a more standard product. Designers can also specify special glass coatings and employ solar shading to prevent overheating, but there is a limit to how much we can improve the thermal performance of a transparent window made of glass.

BELOW Water House 2.0, a prototype building using WFG technology at Feng Chia University.

'Our panel can heat and cool the building itself – the water inside the panel does the very same job as heating… It saves energy, when you compare it to a similar building with large glass surfaces – it's a very clean and sustainable solution.'

Dr Matyas Gutai

Gutai's invention uses an outer layer of water trapped between glass. When sunlight passes through the glass it heats the water rather than the room, preventing internal overheating. When the water is sufficiently warm, it is then pumped into a local insulated storage and thus draws in cooler water. Likewise, the process is reversed in a cool climate when the colder external air temperature is attenuated by the WFG, which when cooled can be pumped away, drawing in warmer heated water. The thin and movable water layer can also absorb the internal heat from a room produced by occupants and equipment.

By using WFG in conjunction with a water source heat pump it is easy to imagine how considerable heat energy (usable for heating and cooling) can be produced with such a system. Gutai also addresses potential drawbacks to this technology such as the water freezing (addition of non-toxic anti-freeze) or pumping out the liquid if necessary. Gutai's technology only requires small energy input to power pumps, which is more than offset by heating and cooling savings. Other benefits include the use of clear (non-tinted) glazing and reduced requirement for external or internal solar shading. The addition of water adds mass to the glazing, which can also mitigate acoustic transmission. Gutai estimates a heating saving of 47–72% compared to a comparable double-glazed window, and 34–61% to a comparable triple-glazed window.

ABOVE A thin layer of water sits between the panes of glass, absorbing solar heat from windows and so reducing the need for active cooling.

2.6 Sand

'A 3D-printed installation in sand and furan resin (cellulose of pine trees and corn kernels), the lattice thickness is based on structural forces and connects visually to the pattern of palm trees, forming a homogenous, natural appearance.'

Chris Precht, Studio Precht

The Sandwaves

Mamou-Mani Architects and Studio Precht

Commissioned to create a temporary installation in Diriyah, Riyadh, Saudi Arabia (an area known for its traditional mud brick architecture), designers Arthur Mamou-Mani and Chris Precht wanted to use the sand that surrounded the site and combine it with a biodegradable adhesive to create a printable composite material. Using an adapted version of technology used in metal sand-casting (to produce moulds), Mamou-Mani and Precht designed a modular system of overlaid branching elements to create panels of dense frameworks made from solidified sand. The installation consists of 58 3D-printed panels that form the backs and the brise-soleil shade for snaking benches. As the installation was temporary the designers were keen to use material that is 100% recyclable, and so they used sand with furan as the resin binder. Furan is a cellulose binder made from pine needles and corn kernels. The 3D-printing

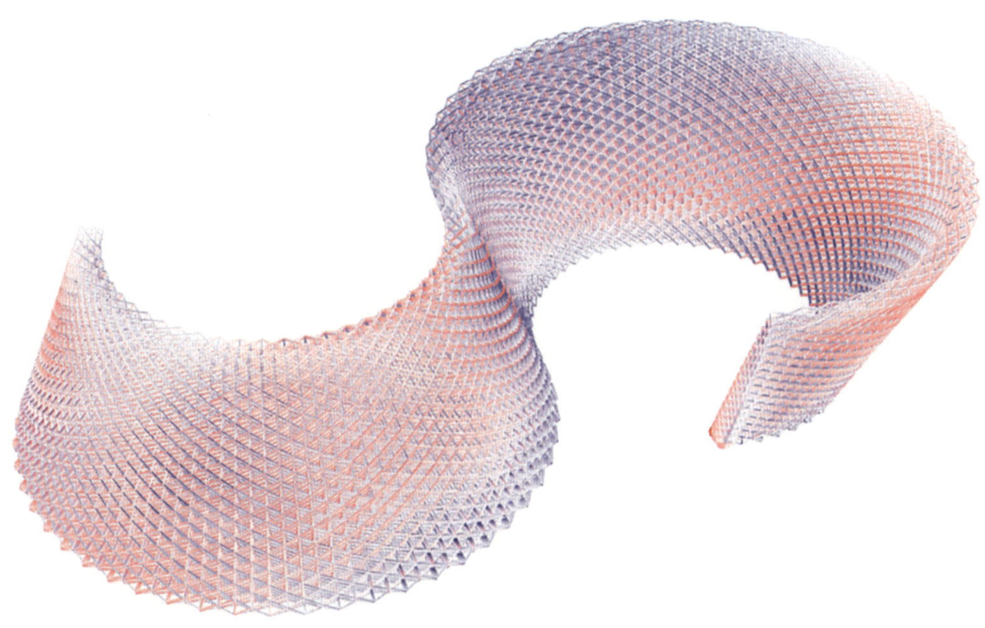

ABOVE Finite Element Analysis (FEA) of structural forces in The Sandwaves installation.

and recyclable polymers. With the printed clay being able to be fired in the same way as any ceramic, this waste-free fabrication method can just as easily be used to print a cup as a building component, and we are now perhaps at a moment when the promise of 3D printing can be more fully realized in the context of the built environment.

While The Sandwaves is a temporary installation as part of a design festival, it represents the kind of speculative endeavour that forms a proof of concept for new ways of making. Likewise, the practice of Mamou-Mani may represent a new model of design office which, though digitally centred, is directly engaged with the physical stuff of material, its quantities, properties, and circular economies.

process builds up thin layers of sand on which the binder is applied, hardens that area and the process is repeated. This additive process allows the printed forms to be structurally supported during fabrication, and then they are removed from the non-solidified sand mould once complete.

The installation was designed to be temporary and was crushed back into sand once the design festival was finished. Mamou-Mani Architects is an architectural design practice that also specializes in digital design and fabrication, and they prototype and fabricate construction elements in their own facility. As well as machining and cutting sheet materials like plywood, they also 3D print with clay

ABOVE When the 3D-printing process is complete, the excess loose and unbonded sand is removed to reveal the structural sand and resin lattice frames.
RIGHT The 3D-printed modular lattice components are unpacked and ready for assembly of the final installation.

CHAPTER THREE

Grown Materials

Grown materials

Why don't we farm the materials for the construction of our buildings? To an extent we already do, and globally recognized schemes such as the Forest Stewardship Council (FSC) have successfully promoted sustainable forestry and the prevention of deforestation. Timber is, with well-managed forests, a sustainable and carbon-negative construction material, absorbing (or sequestering) carbon during growth. Less environmentally palatable are the fungicides and pesticides used to preserve and prolong the serviceable life of timber. Non-chemical treatments of non-durable timber are explored and the Life Cycle Analysis (LCA) of cross-laminated timber (CLT) is brilliantly illustrated by engineered timber advocates Waugh Thistleton (p.121). In Elding Oscarson's beautiful new Wisdome building in Stockholm (p.124) the architects show off the structural properties of a timber gridshell, whereas Facit Homes demonstrate the lightweight strength and ease of assembly of their cut to order plywood sheet housing system (p.131). At De Nieuwe Veemarkt housing development in the Netherlands (p.128), Studio Nauta creates a housing complex where the entire superstructure, foundations, insulation and roofing are bio-based. The Cork House (p.136) remains a useful outlier, showcasing the use of cork as a singular substrate that acts as both structure and environmental envelope.

'All of our materials, products and biotechnologies are designed with circularity at heart. We are driven by a very simple philosophy: to allow nature to lead innovation and only have a positive or regenerative impact on everything we touch.'

Dr Ehab Sayed, Biohm

Paper and paper pulp are most often made from the wood fibre waste of the timber industry. This cellulose-based material is shown as a type of paper concrete in HiLo Lab's experimental Hydrox Wall (p.139), and a dense gypsum and paper-machined lining to Herzog & de Meuron's Elbphilharmonie in Hamburg (p.141). The Wikkelhouse by Fiction Factory (p.144) uses corrugated cardboard to form the structural skin of this brilliant prefab housing module, and researchers at TU Darmstadt in Germany explore the technical limits of multi-layered paper building systems (p.148).

The fast-growing materials of hemp, bamboo and other grasses feature in the work of material specialists like William Stanwix (hemp, p.151) and Jan Balbaligo (bamboo, p.156), with their material knowledge gained through innovative construction projects and direct hands-on experience. New timber and bamboo composites are explored by professor Hexin (Johnson) Zhang of Edinburgh Napier University (p.159), and the sugar cane waste of an East London sugar refinery is brilliantly explored as a viable building material by staff and students from the University of East London in collaboration with Grimshaw architects (p.162).

Mycelium is explored as a 3D-printed material by Blast Studio (p.176), and as a structural component in Dirk Hebel and Philippe Block's MycoTree pavilion (p.178). The fabrication process or cultivation of the mycelium acoustic panels and flooring of SQIM and Mogu (p.173) highlight how waste streams are carefully transformed and grown into moulds in a beautifully documented process that is more akin to a laboratory than a workshop.

Two very different types of algae architecture are illustrated with Prometheus Materials (p.181) creating a new type of bio-cement to produce a sustainable alternative to the concrete block, while ecoLogicStudio (p.183) incorporates algae bio-reactors in the walls of an inflatable pavilion to clean air and suggest a different kind of biogenic architecture. Paulo Gomes invents a way to add economic and social value to the recycling of defunct plastic garden furniture in his Transformation Workshop project in Brazil (p.187), and the recycling of scallop shells into a bio-composite in the form of the 'Shellmet' by TBWA\Hakuhodo + Quantum (p.190) is a neat and inventive example of Japanese circular economy. This grown materials chapter ends with salt, and the experimental work of Karlijn Sibbel and Henna Burney of Atelier LUMA (p.199). As part of the ongoing construction of the Luma Foundation in Arles, panels of salt have been grown in the saltwater ponds of the Camargue and used as an internal building finish, which is both visually appealing and very good for air quality.

OPPOSITE Abundant eelgrass collected from local meadows is packed into cylindrical nets and utilized as insulation for Vandkunsten Architects' Modern Seaweed House (p.169).

3.1 Timber & Cork

'The Confederation of Forest Industries (Confor) recently warned that the UK faces declining supplies of homegrown wood due to a lack of productive tree planting. With the country needing to import more than 80% of its wood requirement, the UK could be sleepwalking into a timber shortage in the not-too-distant future. We need to urgently move productive tree planting up the agenda.'

Stuart Goodall, CEO of Confor

Non-chemical timber treatments

Accoya and Vastern Timber

If sourced from a sustainably managed forest, timber is an excellent low-carbon construction material. However, to make softwood timber durable the wood is often treated with harmful chemical preservatives. Problems with wood preservatives can include preventing its sustainable reuse, as the chemicals can be released into the atmosphere or leach into the water course. These chemical timber preservation treatments are designed to repel and kill fungi and insects like beetles, and as such their toxicity is non-selective and can be harmful to people as well as the larger environment. With the whole life of materials, including potential reuse and eventual disposal, now being taken seriously, new and less toxic alternatives to preserving the structural integrity of timber are being employed.

BELOW Softwood timber being loaded into a huge kiln for thermal modification to preserve and protect the timber against insects and rot.

Accoya is not a type of tree species but is the name of a patented timber preservation treatment. The technology of Accoya is based on wood acetylation. Acetylation changes the free hydroxyls within the timber into acetyl groups with acetic anhydride, deriving from acetic acid (known as vinegar in its dilute form). When the free hydroxyl group is transformed to an acetyl group, the ability of wood to absorb water is reduced by almost 80%, helping the wood to be more dimensionally stable, preventing rot and insect ingress, and thus creating a more durable and non-toxic material. Accoya timber is usually made from a sustainably sourced 30-year-old pine tree. The process of acetylation transforms the whole timber element and is not simply an exterior coating or protection, so the treated Accoya timber can be cut and machined and is good for making timber windows, doors and building cladding.

The technology to thermally modify timber was commercialized in the 1990s, and there is increasing interest in how we can extend the longevity of softwood timber without the use of harmful fungicides and pesticides. The thermal modification process uses steam and heat of 190°C (374°F) within a specially constructed kiln. The process not only reduces the water content but also changes the cellular structure of the timber, reducing the ability of the timber to absorb water and thus becoming less prone to rot. The thermal modification process also creates a more dimensionally stable timber. The process of thermally modifying timber does use energy, and as designers and specifiers of materials and products it is still within our gift to properly consider the technical requirements of any given project and the appropriate sourcing and processing of timber.

Depending on your site location are there opportunities to use more local material? In Austria, where almost half the country is covered in well-managed forests, there is a well-developed timber, sawn timber and engineered timber industry and so you do not have to go far for supplies of high-quality softwood timber. In contrast the UK imports an estimated 67% of its sawn timber, and a more sustained approach to productive softwood tree planting (which accounts for 95% of UK construction timber use) as well as better use of UK-grown and thermally modified hardwoods like Vastern Timber's Brimstone products, could obviate the use of chemical timber treatments.

BELOW The processes of thermal modification and acetalization create a more dimensionally stable construction material without the use of harmful pesticides and fungicides.

Black & White Building

**Waugh Thistleton Architects
Location: London, UK**

KEY DATA

Upfront embodied carbon: 329
$kgCO_2e/m^2$

Whole life embodied carbon: 528
$kgCO_2e/m^2$

BELOW The structure of the Black & White Building consists of a beech laminated veneer lumber frame with softwood CLT floors, walls and core; steel structural connections are buried in the timber elements.

The Black & White Building is the tallest engineered timber office building in central London. Built on a constrained site in Shoreditch, the building is a perfect case study for the use of contemporary timber technology and minimizing embodied carbon and operational energy. Waugh Thistleton Architects (WTA) have pioneered the use of engineered timber products such as cross-laminated timber (CLT) in the UK, and their commitment to creating a new sustainable design and construction methodology has become their USP. Previous WTA projects such as Pitfield Street and Dalston Works have used CLT as the structure for walls and floors, and their Orsman Road project used a hybrid of a light steelwork frame with CLT floors, walls and roof. The Black & White Building is the first project where the practice uses a combination of CLT for floor slabs, walls and core, and laminated veneer lumber (LVL) for columns and beams. The LVL components are visibly more engineered, denser and stronger and made from a laminate of thin beech hardwood timber veneers, whereas the CLT components are lighter and simpler slabs of cross-laminated thicker softwood spruce components.

'It is a massively carbon-saving building that comes way below the targets set out by London Energy Transformation Initiative (LETI).'

Andrew Waugh, Waugh Thistleton Architects

A key part of the WTA design process for the building was the Life Cycle Assessment (LCA), which designs for the whole working life of the building and beyond, with the LCA calculated with a 60-year building life. As such the building is designed to be disassembled and the building frame is made as a series of pre-fabricated elements that are assembled and bolted together on site. Construction images clearly show how this building process is precise, straightforward and clean, with minimal 'wet trades' required on site. The timber components are connected with steel 'flitch' plates cut into the timber members and then essentially wrapped in timber, bolted and protected from fire. The bolted connections allow for ease of disassembly and potential reuse of the structural members. The engineered timber structure has 50% less embodied carbon than the equivalent reinforced concrete frame. Using prefabricated timber elements also reduces construction time, lowering associated onsite energy requirements and reducing site deliveries by 85%: a single load can deliver 40m³ (1412 cubic ft) of timber versus a single load of concrete which can only deliver 6m³ (212 cubic ft).

The glazing system is made of glue-laminated timber frames as opposed to higher carbon aluminium or steel alternatives. Detailed environmental design, modelling and analysis has led to the incorporation of key passive design features such as natural cross ventilation and solar shading to mitigate the operational energy/carbon needed to heat and cool the building. The external louvres are vertically oriented on the north and west façades and horizontal on the south-facing façade. Six different depths of louvres are employed over the building, and they are manufactured from fast-growing American tulipwood, which has been thermally modified (a non-chemical timber-preservation treatment). Interestingly, WTA's own Life Cycle Assessment report highlights (in a 'lessons learnt' section) how the single-storey reinforced concrete basement accounts for a massive 46% of upfront embodied energy for the project.

LEFT There are no structural walls other than the stair and lift core, thus allowing maximum flexibility for future use.

OPPOSITE Solar shading is provided via timber louvres, minimizing the use of reflective glass coatings and maintaining good daylighting while preventing unwanted solar thermal gain.

Wisdome: timber gridshell in laminated veneer lumber (LVL)

Elding Oscarson, Florian Kosche, Blumer Lehmann AG, Design-to-Production, Stora Enso Location: Tekniska Museet, Stockholm, Sweden

KEY DATA

Roof dimensions: 48 x 26m (157½ x 85½ ft)

Dome dimensions: 21.6m diameter x 12.2m high (71 x 40 ft)

Embodied carbon: Manufacturers Stora Enso estimate that 132 tonnes (145 US tons) of greenhouse gas emissions were emitted during the manufacturing of the LVL, with 683 tonnes (753 US tons) of carbon dioxide removed from the atmosphere when the trees were growing and stored in the wood over the lifetime of the Wisdome project.

Fabrication process: Bonded layers of spruce timber veneer, laminated into gridshell beam system.

The Wisdome project is the extension to the National Museum of Science and Technology in Stockholm, Sweden, and is in itself a wonderful exhibition of contemporary design, fabrication and construction technology. The rectangular site is entirely covered by a timber gridshell held off the ground by edge columns at 6m (20 ft) centres. In the middle of the covered space a free-standing geodesic timber dome houses a 100-seat cinema for 3D film projection. The principle of the sustainable timber gridshell was developed by engineer Frei Otto and used to great effect in the Multihalle in Mannheim (1975). Large spanning and lightweight roofs can be created by producing doubly curved two-way timber lathes that utilize the flexural strength of timber and minimize material use.

The gridshell is fabricated from five alternating layered beams 155mm (6 in) deep, with each beam

---

OPPOSITE A cross-sectional model shows the layering of the various timber 'skins'; the exterior of the timber gridshell is covered with timber shingles (or 'shakes').

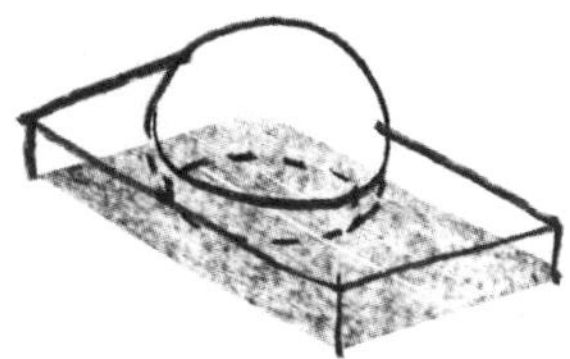 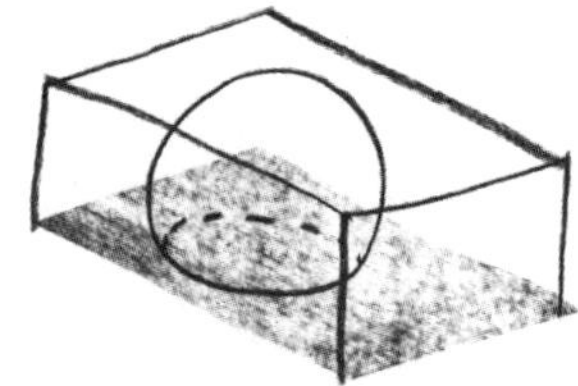 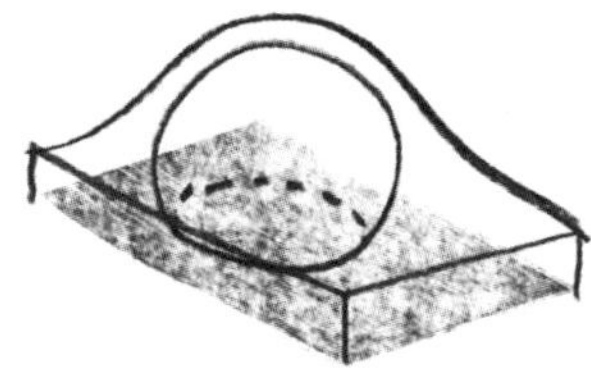

ABOVE Architects' concept sketch: a laminated timber gridshell roof wraps around a geodesic timber theatre made from cross-laminated timber.

'There are plenty of cars, machines and aeroplanes in the museum, but there is little about construction.'

Johan Oscarson, Elding Oscarson

consisting of five layers of 31mm (1¼ in) thick laminated veneer lumber (LVL), with three beams spanning the short section and two spanning the long section. The layers work together as a single gridshell structure through a series of profiled timber dowels that transfer load and shear forces through the structural framework. Having tested a mock-up fabricated section at full scale it was decided to pre-laminate the bottom beam, which then acted as falsework for the rest of the structure. The subsequent beams were built up in 31mm (1¼ in) thick planks of LVL on site, where they were glued and screwed together. This allowed

for ease of road transportation as the individual elements could be broken down to transportable-sized lengths of no more than 13.3m (44 ft). The layering of the LVL uses staggered joints in order to maintain smooth overall double curvature for each beam. The specific LVL used in the Wisdome project is made from multiple veneers of strength-graded dried spruce bonded together in Stora Enso's factory in Varkaus, Finland.

This intricately fabricated structural roofing system is the product of a successful collaboration between designer, engineer, digital modelling/fabrication and material supplier. The roof structure is covered with a solid layer of LVL, laid as diagonal strips, with insulation, waterproof membrane and an outer visual skin of timber shingles (or shakes).

The timber gridshell roof is supported at the edge by 24 solid timber columns made from laminated LVL with dimensions of 600 x 800mm (2 x 2 ½ ft) and 600 x 600mm (2 x 2 ft). The columns have a rigid connection to the concrete foundation and in order to resist the lateral (sideways) force of the roof, steel tension rods run through the timber columns, which are pre-tensioned after assembly. A hollow box 'edge beam' sits atop the column heads, connecting roof and support columns.

LEFT Detail of the timber gridshell shows five alternating layers of laminated timber elements made from Finnish spruce.
OPPOSITE Visualization showing the contrasting timber technologies of laminated gridshell roof and cross-laminated timber geodesic theatre dome.

cafe dome

De Nieuwe Veemarkt

**Studio Nauta with Mulder Zonderland, Schipper Bosch, Joost Emmerik, BC Materials, And The People, Treetek, Econu, Solid Timber
Location: Zwolle, the Netherlands**

Studio Nauta was selected from over 50 entrants to design a unique mixed-use housing development in the municipality of Zwolle in the northeast of the Netherlands. De Nieuwe Veemarkt is described as a competition to design 130 bio-based homes (mostly social and affordable) as a holistic community including energy production, car and bicycle storage, and built in and around a community garden.

Part of the strategy of the development is to challenge the orthodoxy around the construction process and associated supply chains, and take a hyperlocal approach that aims to source materials, suppliers and labour proximal to the project, with a radius of 100km (62 miles) set as an ambition. Studio Nauta is working with Belgium BC Materials, a cooperative founded in October 2018 with the mission to promote earth construction as one of the most important solutions to counter the enormous environmental impact of the construction sector.

____

BELOW The housing development is built on the site of a former cattle market and consists of a series of inter-connected housing blocks of between three and five storeys.

'Substituting carbon-intensive technical materials with regenerative resources and materials from the biosphere, which absorb and store natural carbon, has become a key approach to decarbonizing our built environment. The techniques and technologies for bio-based manufacturing and construction are well established, but the infrastructure and frameworks are not established in scale to support them.'

Arup and Materials Cultures

The main structure is a glue-laminated timber (glulam) frame with cross-laminated timber (CLT) cross walls and an innovative timber and rammed earth flooring system devised by Swiss construction innovators Rematter (see p.36), with the earth sourced locally from excavation material. External walls are insulated with a thick layer of flax insulation between the glulam columns, and the exterior is clad with untreated larch timber shingles. Additional thermal and acoustic insulation between apartments is supplied by sprayed-in cellulose cavity wall insulation made from timber and paper waste. The foundations of the buildings are provided by spruce wood piles sourced from the Ardennes. These timber poles are available in lengths from between 1 and 23m (3¼–75 ft) in

BELOW Drawings show sections of the flat roof and residential wall, and the floor with façade and balcony, listing the impressively sustainable materials palette.

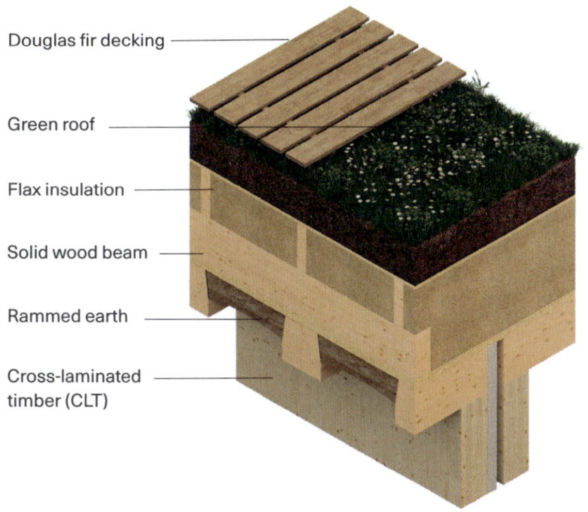

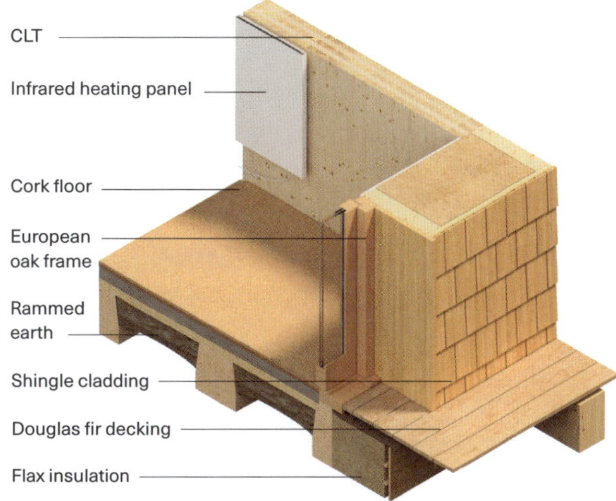

length, and they are driven into the ground with lightweight plant equipment. Importantly, the timber piles must be fully submerged in the ground to prevent rot, with the top of the timber pile at least 200mm (8 in) from the ground plane, so the piles are capped with small concrete caps that form the connection between the glulam timber frame above. While this building is not entirely concrete free, the use of this high-carbon substrate is extremely limited for a building of this size. Other innovative bio-based materials used in the project are the floor insulating and levelling compound made from wood chip and minerals and then floored with cork. The green roof system is also waterproofed with a non-petroleum-based asphalt alternative made from a combination of sugar waste, rice, corn and potato starches, natural tree and gum resins, natural latex rubber, vegetable oils and lignin. As well as showcasing a thoughtful approach to new housing projects, projects like De Nieuwe Veemarkt also present an invaluable case study in aspirational bio-based construction technology that supports local industry, jobs, skills and the natural environment.

BELOW The integrated plan includes 130 homes, a community centre, an energy production hub and car and bicycle storage, and is organized around a communal garden where people, animals and plants live in harmony.

Facit Homes: digitally manufactured timber house

Bruce Bell

KEY DATA

Embodied carbon: 5933 kgCO$_2$e for 2-storey Facit House as opposed to 49,062 kgCO$_2$e for a 2-storey masonry house. These figures are calculated based on a house of two floors with usable floor area of 80m^2 and typical U-values of 0.14 W/m^{2}K for all elements.

BELOW Both complex and more standard chassis components made from plywood sheets can easily be fabricated using the Facit Homes construction approach.

As a founding partner of Facit Homes, industrial designer Bruce Bell wanted to harness the power of new (and affordable) digital technology to improve how new homes are designed, built and technically perform. The Facit Home uses FSC-certified 18mm (¾ in) thick spruce plywood sheets, CNC machined to 0.2mm tolerance and fabricated in 300mm (12 in) deep boxes which provide structural integrity as well as deep voids for thermal insulation. Lightweight timber composite I-beams are used for key spanning elements with specially fabricated 'Facit' components spanning doors and window openings. The plywood components are cut and fabricated within Facit's mobile production facility, which is housed in a single shipping container delivered to site at the start of the construction process. All plywood is delivered directly to site during construction, with waste offcuts collected by the supplier and returned

'We make it work in our own way, there is not anyone in our company who is from a construction background – we are engineers, architects, technology managers, financial people. We don't really look like a contractor, yet we build homes.'

Bruce Bell, Facit Homes

for recycling as paper products. However, waste is minimized with the use of Facit's own BIM (building information modelling) platform, with all sheets optimized and components 'nested' onto sheets, thus shortening cutting time and mitigating offcuts and associated waste.

After completion of the cutting, assembly and placement of the timber cassettes, thermal insulation is added to the hollow-box wall through pre-cut openings. Facit is currently using 280mm (11 in) of pressure-injected expanded polystyrene (EPS) beads. These provide a U-value of 0.12W/m²K and do not produce any waste, as you spray as much as you need to fill the wall cavities. Facit has also previously used a sprayed-in cellulose insulation made from waste newspaper pulp.

Bell has written and commented extensively on the UK modular timber construction industry and its focus on off-site construction. Good arguments are made around the improved quality that comes from a protected factory environment, but the shipping of large-scale and heavy buildings (or at least parts of building) around the country seem to be counter-productive, and the high-profile failure of some of these modular construction firms either highlights an industry resistant to change or confirms that this model is not appropriate for the UK. In Japan and the US there are well-established prefabricated building industries and associated supply chains, and the success of designs like the Wikkelhouse (see p.144) in the Netherlands demonstrates that

specific building processes are only as successful as the construction culture and infrastructure in any particular region or country.

The quiet but brilliant innovation of the Facit system is not born out of designer dogma or obsession with technology for its own sake; it stems from a genuine interest in harnessing the usefulness of digital design and fabrication in the construction of sustainable housing with ultra-low embodied carbon and minimized operational energy demands through a high-performance building envelope. The Facit system is open, so each of their projects is not and does not look the same; and while Facit uses a determined building skin depth, its system is not modular and does not need to produce standard panel components. The system also challenges the idea of centralized manufacturing. It embraces a new model that Facit describes as de-centralized digital manufacture, which includes ideas of manufacturing on site and on demand, minimizing associated shipping and logistics, lowering factory overheads and improving job satisfaction.

OPPOSITE, ABOVE Plywood sections of walls and floors can be easily assembled by hand. Facit Homes' mobile production facility at the rear contains a large CNC machine to cut timber sheets.
RIGHT The entire structural carcass of the house, including the floors, walls and roof, can be fabricated from 18mm (¾ in) thick plywood sheets.

Timber 'tree fork' connections

Caitlin Mueller, Building Technology Program, MIT Energy Initiative

Caitlin Mueller, associate professor of architecture and civil and environmental engineering on MIT's Building Technology Program, identified an opportunity for using natural structures in timber construction. The waste products of engineered timber include irregular sections such as knots and forks that are typically turned into pellets and burned or chipped for garden mulch. Mueller and her Digital Structures research group have devised a strategy for the reuse or upcycling of these timber elements as valuable structural components.

In the race to make construction ecologically sustainable there is an increasing demand to replace high-carbon materials such as concrete or steel with engineered timber. Engineering timber to create columns, beams, floor and wall panels requires a uniformity of timber which commercially farmed conifer trees such as pine, spruce and larch

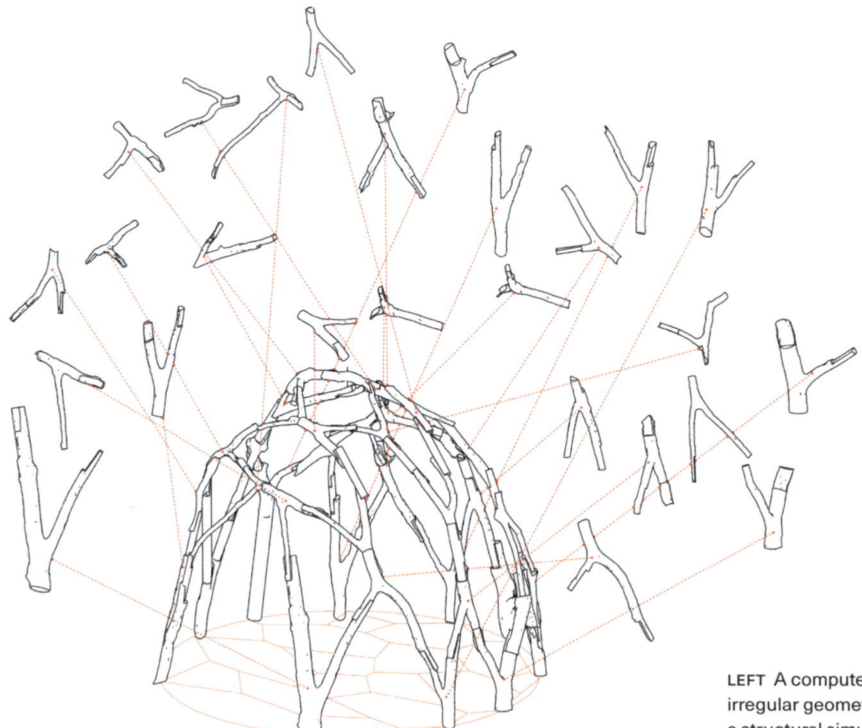

LEFT A computer algorithm can match the irregular geometries of the tree forks into a structural simulation for design iteration.

'Tree forks are naturally engineered structural connections that work as cantilevers in trees, which means that they have the potential to transfer force very efficiently thanks to their internal fibre structure.'

Caitlin Mueller, Department of Architecture, MIT

can provide. This processing of timber into thin veneers for LVL or thicker strips for CLT utilizes the composite strength of the timber as a material, but not the innate strength of branching joints formed through tree growth.

Caitlin Mueller's brilliantly simple idea is to make use of natural structures in timber construction and use off-cut tree forks as load-bearing joints. So rather than treat these tree forks as an inconvenience and low-value waste product, they are employed as high-strength and high-value structural connections. As trees naturally bifurcate and form branching systems the tree wood hardens around these junctions to resist the cantilevers of the branch. Having analysed the fibre orientation and density of tree forks, Mueller's MIT team has identified that these Y-shaped components can be re-deployed as new structural connections for building.

Mueller and her team have subsequently developed a design-to-fabrication workflow sequence to successfully integrate the tree fork components into a larger digital fabrication design process. This process includes identifying a suitable source of timber, which Mueller did by working with a municipal forestry group in Massachusetts, and when a large group of trees was felled for the construction of a new school in Somerville, waste wood was diverted to Mueller's programme. The team then created a digital library of the tree forks using photogrammetry and other low-cost 3D-scanning techniques. This library of components is then matched to suitable connections of a designed structure using a simple matchmaking programme known as the Hungarian

algorithm. Machine code is then generated for the fast cutting of the tree forks to provide good longitudinal connections to other structural members, and finally the structure is assembled.

The finished building is designed to include approximately 40 nodes and will be installed as an outdoor pavilion on the site of the previously felled trees in Somerville.

BELOW Mueller and her team installed their first prototype gridshell structure, which incorporated 12 tree fork nodes, on the MIT campus.

Cork House

**Matthew Barnett Howland,
Oliver Wilton and Dido Milne
Location: Eton, Berkshire, UK**

KEY DATA

Upfront embodied carbon:
−18 kgCO$_2$eq/m^2

Whole life embodied carbon:
618 kgCO$_2$e/m^2

Overall building envelope U-value:
0.15 W/m^{2}k

Cork House is a world first, a permanent building with solid structural cork walls and roof, designed to fully meet relevant UK building regulations. The completion of Cork House also successfully marked the end of a research project started in 2014 by architect Matthew Barnett Howland and architect/researcher Oliver Wilton. Barnett Howland and Wilton wanted to create a low- or no-carbon plant-based construction technology that utilized the natural properties of waste cork bark.

Cork is not a new construction material, with evidence of its use in the ancient Nuragic civilization of Sardinia, and more recently for flooring and thermal insulation in the 1930s.

____

BELOW Steel rooflights are fixed to an Accoya frame atop a cork block corbelled roof. The rest of the house is made from solid cork block walls, with Accoya beams spanning structural openings.

'We used it (cork) to fulfil all the functions that you need a wall and roof to do – structure, insulation, the external finish, the internal finish – it does the whole lot, so it was a really simple construction.'

Matthew Barnett Howland

Cork forms the outer bark of the cork oak tree, typically grown in the Western Mediterranean basin, notably Portugal. It is harvested every nine years, gently stripped from the trees without damaging them, and is then used (in its unprocessed state) for high-quality wine bottle stoppers. The waste cork from this process is granulated and used as cheaper wine corks, and as other industrial products with the addition of chemical binders.

The architects were interested in using a pure expanded cork (without chemical additives), which is formed by heating cork granules in autoclaves using super-heated steam. This process causes the granules to expand, blacken and bind together, with the heat process causing the natural suberin resin in the cork to be released and bind the granules together to form expanded cork blocks or slabs. Amorim, the company that supplied the cork for the research project and house, manufacture with 90% biomass energy and the Environmental Product Declaration (EPD) for the cork classifies the material as carbon negative, in that it sequesters or stores more carbon than it will produce over its lifetime.

Most of the building envelope (excluding the floor) is made from expanded cork blocks that provide the structure, thermal insulation and weather-tightness of the house. This was achieved through a rigorous prototyping and testing process in relation to fire, driving rain and airtightness. Working with the Bartlett (UCL) fabrication lab, blocks were CNC machined to interlock, allowing the whole building to be dry-jointed, avoiding adhesives, minimizing mechanical fixings, and providing for future demountability. The house comprises 1268 interlocking machined cork blocks. The door, window frames and beams are made of Accoya acetylated pine timber, and the roof includes cedar weather boarding after extensive testing proved this was required to repel driving rain. The distinctive shape of the building allows the wall construction to turn into the roof construction by employing the ancient technology of corbelling to form the five square pyramidal roofs. Cork House features a water sprinkler system as the designers did not want to internally clad and cover the cork walls or treat the walls with chemical fire retardants.

BELOW The cork blocks sit on a CLT deck supported by steel screw piles, providing a cement-free building foundation.

3.2 Paper

'Paper materials have good thermal insulation properties. The thermal conductivity of paper materials is already comparable with conventional thermal insulation materials without any additional optimization processes. Recycled cellulose is commercially available as insulation material using the blow-in or pour-in process.'

Rebecca Bach, Alexander Wolf, Martin Wilfinger, Nihat Kiziltoprak, Ulrich Knaack (TU Darmstadt)

Hydrox Wall: paper pulp concrete

Blair Satterfield, Marc Swackhamer
(LoDo Lab), Jason Heinrich, Neal Li and
Thomas Gaudin (HiLo Lab, University of
British Columbia); HouMinn Practice

The HiLo Lab is based at the School of Architecture
+ Landscape Architecture, University of British
Columbia (UBC). The lab is so named because it aims
to work with lowly (sometimes designated waste)
materials in highly sophisticated design applications.
HiLo also usefully describes a design approach that
embraces both the high-tech of digital fabrication
with the low-tech sophistication of tacit construction
knowledge. This academic research lab works
at the intersection of materials research, ecology
and the use of novel technologies in architectural
applications. The lab's mission centres on three
interrelated ideas: the use of second stream (waste)
materials in construction; providing increased access
to digital design and fabrication processes; and the
design and application of energy-efficient methods

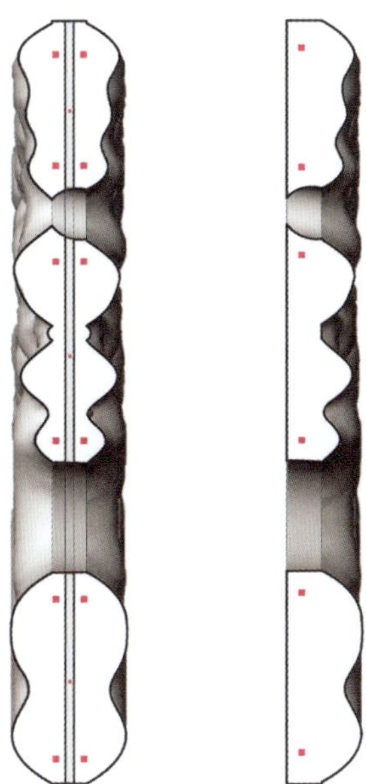

ABOVE Drawings show the structural 'sandwich' of fabric formed
by paper pulp outer skins bonded with a magnesium oxide concrete core. Bamboo reinforcement is indicated by the pink dots.
RIGHT Sample of the paper pulp substrate cast in fabric formwork.

for construction that minimize or eliminate waste. The paper pulp concrete 'Hydrox Wall' was a research project, designed as a prototypical internal wall panel for providing acoustic separation or at least attenuation between two ends of a university studio space. Paper pulp was produced from harvested office waste from the campus, and was combined with a magnesium oxide concrete core to form a structural laminate wall. Importantly, magnesium oxide binds with cellulose (paper pulp), unlike Portland cement.

Each panel was formed horizontally in two pieces by creating light timber frameworks with attached fabric formwork. The panels were then filled with a paper pulp mixture, with bamboo rods used as tensile reinforcement. Once dry these panels were lifted and positioned as formwork for a poured magnesium oxide concrete and paper pulp core that binds both surfaces together, creating a strong sandwich-like structure. The Hydrox name comes from an old North American cookie made from two outer layers with a cream filling.

The HiLo lab worked with UBC Civil Engineering to test the compressive strength of the concrete and paper pulp mix and the relative usability of the different mixtures. Ultimately HiLo created surfaces out of paper (with magnesium oxide concrete as a binder), and complete panels out of a sandwich of these panels with concrete and paper pulp filler.

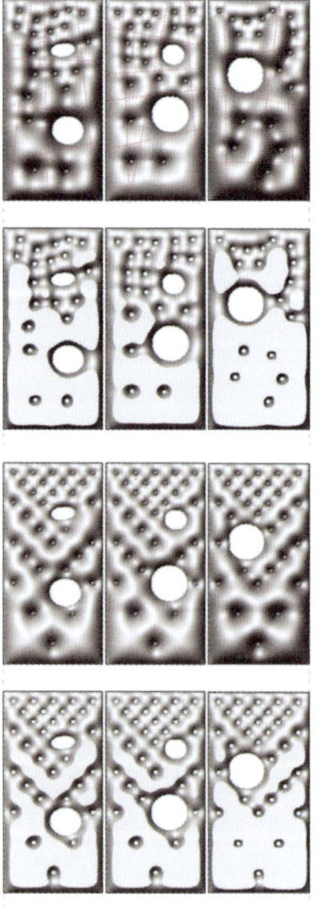

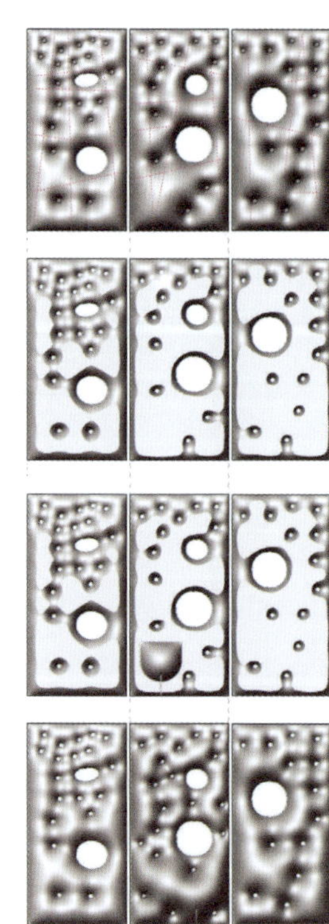

BELOW The Hydrox sandwich of paper pulp outer skins with an internal magnesium oxide concrete core bonding them together; iterative design options looking at visual transparency and structural stability.

The Elbphilharmonie: paper and gypsum 'white skin'

**Herzog & de Meuron, Nagata Acoustics, Knauf
Location: Hamburg, Germany**

The extraordinary Elbphilharmonie in Hamburg sits atop a former brick cocoa warehouse on the Hamburg harbour facing the river Elbe. The iridescent glass box with lifted edges forms a gravity-defying crown-like form. The entire building complex includes housing and a luxury hotel, and at its centre a concert hall for over 2000 people.

The 10,287 wall panels that line the concert hall were specially designed by the architects in conjunction with acoustician Yasuhisa Toyota, and were 'acoustically microshaped' to create the correct amount of sound reflection and absorption. The 'white skin' panels were manufactured by Knauf and are made from a mixture of natural gypsum plaster and recycled paper pulp fibres. Up to five sheets of 36mm (1½ in) board were bonded together then machined at depths of between 5 and 90mm (¼–3½ in) to create the unique interior surface morphology. The panels provide a highly tuned acoustic environment as well as fire protection to the auditorium.

The combination of gypsum plaster and paper as a construction material is not new, and has been used since Augustine Sackett and Fred L. Kane invented plasterboard (drywall) in 1890. Plasterboard is a layer of gypsum plaster encased in a paper skin, either side of which provides enough tensile strength to prevent the boards cracking or disintegrating before installation. Plasterboard provides excellent fire protection for structures and is 100% recyclable, although it is still not widely recycled.

BELOW The paper pulp panels of the auditorium have been individually machined to create the appropriate acoustic for a concert orchestra.

'The architect favoured the seashell motif microshaping for the diffusing surface pattern, which were usually expected to create soft sound reflections. They were also used to eliminate the long path echoes. The depth of microshaping at each portion of the wall and ceiling was defined through the acoustical test with a 1:10 scale physical model.'

Yasuhisa Toyota, Nagata Acoustics

The Elbphilharmonie internal cladding panels are a composite mixture of recycled paper fibres and gypsum plaster, creating a homogenous and strong matrix which can then be machined with a CNC cutter. Gypsum is a dense, heavy material, and so the layered-up panels of the concert hall lining have a mass useful for the attenuation of sound. Gypsum has been used as a construction material for thousands of years, with gypsum found on the interiors of the great pyramids in Egypt. Gypsum mining is seen as relatively low-impact and the heat treating it requires during processing is no more than the temperature of a domestic oven, and thus relatively low-carbon. Gypsum plasterboard can be recycled; the outer paper facer is removed from the gypsum board, and the gypsum core can be recycled as an additive to concrete, plaster and render. The core material can also be reused as a soil amendment to improve water retention, and added to animal bedding to absorb moisture.

OPPOSITE Almost the entire interior of the auditorium is finished in the paper and gypsum panels, which also give the space its distinctive colouring.

BELOW The panels are custom machined to control their acoustic performance in relation to absorption, reflectance and diffusion of sound.

Wikkelhouse: cardboard house

René Snel and Fiction Factory

KEY DATA

Thermal conductivity: 0.24 W/m²k

OPPOSITE To make the house, a jig is slowly rotated and cardboard wrapped around; one version of the house has the cardboard replaced with a flax fibre blanket to provide thermal resistance.
BELOW Modules that comprise the Wikkelhouse.

The Wikkelhouse was invented by René Snel and has subsequently been developed and built by Fiction Factory, a design and production company based in Amsterdam in the Netherlands. It is a modular building system designed for small houses, holiday cabins, garden rooms and house extensions. The Wikkelhouse is made from a laminate of 24 layers of corrugated cardboard wrapped around a house-shaped cross-sectional frame of CNC-cut radiata pine plywood. The delicate plywood framework provides the form and cross-sectional shape of the house and works like the internal structure of an aeroplane wing, with the glued cardboard skin acting together to form a semi-monocoque or stressed skin structure. If you carefully study the cross-sectional profile of the house, you can see that the walls are also slightly curved, which helps with the structural performance of the building and performs like a tube.

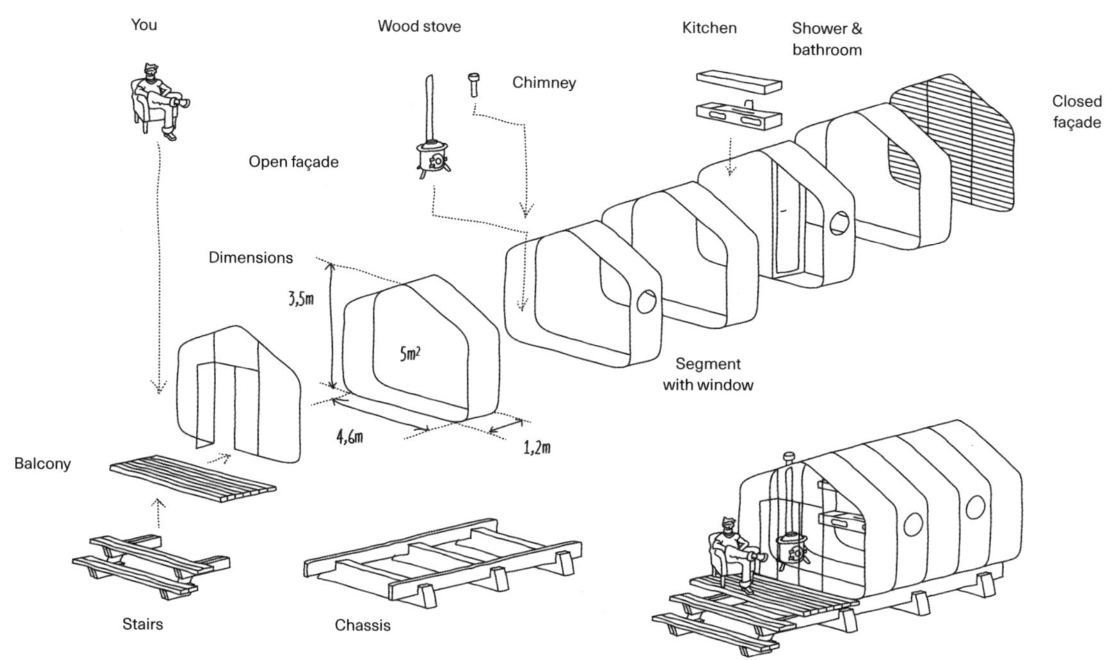

'The project is a house made from cardboard… actually it is a hybrid of plywood and cardboard, because if you only make it from cardboard you run into multiple problems. What is special about the house is that it is light, the cardboard is very strong, it insulates… it is kind of a weird project – it is also intended to inspire.'

Oep Schilling, Fiction Factory

Within the depth of the timber ribs, the building skin is split with a timber connector, which also provides a void for electrical conduit, with 12 layers of cardboard each side providing thermal insulation. There is also a version of the modules where the timber is beefed up and the cardboard is replaced with a flax fibre blanket to provide thermal resistance. 'Wikkeling' is Dutch for wrapping, and that is exactly how these houses are formed. Fiction Factory designed and built a special rotating jig onto which the 1.2m (47 in) deep house section timber frames are loaded. The jig is then slowly rotated and, using low-toxicity or non-toxic adhesives, the cardboard is wrapped around the whole building with the laminating and gluing forming a tough building skin. A waterproof and breathable membrane is then added, along with an exterior timber lathe rainscreen cladding for added durability.

The house is manufactured in 1.2m (47 in) deep bays which are mechanically attached to form the building. Each completed Wikkelhouse bay

or segment weighs 600kg (1,322 lbs), and thus an 8-bay deep Wikkelhouse of 40m² (430 sq ft) will only weigh 4.8 tonnes (5¼ US tons), so carbon-intensive foundations like concrete are not required. The Wikkelhouse is designed to be connected to a simple chassis of two longitudinal timber members which are laid and levelled on the ground. Depending on the site geology and topography of the house, the timber chassis can be held in place by steel screw piles or simple crossbeams with spreader plates. The bays are then connected on site and hooked up to services as appropriate. The interior is lined with a thin plywood skin and services, bathrooms and kitchens can be incorporated into the design. Importantly, the size of the 1200mm (47 in) deep modular building sections mean that they can easily be transported without the logistical issues connected with moving whole houses or anything larger than a shipping container.

As an integral part of the research and development process of the Wikkelhouse, Fiction Factory undertook rigorous structural and material testing of the material and construction of the house. The director of Fiction Factory, Oep Schilling, also commissioned a report from the University of Utrecht specifically studying the relative sustainability of the cardboard and timber used in the house. Schilling wants to be able to legitimately state that Wikkelhouse is a sustainable alternative to other forms of housing.

OPPOSITE, ABOVE 'Wikkelboats', which utilize the structures as house boats on floating pontoons in Rotterdam's marina.
LEFT Finished modular sections of the Wikkelhouse show how thin the structural walls are. The sides and corners are curved to increase structural stiffness.

'Circular' exterior wall system made of paper-based materials

Ulrich Knaack, Nadja Bishara, Inés Burdiles, Naomi Bosse (Institute of Structural Mechanics and Design [ISM+D] Technical University of Darmstadt) with project partners IBC, MIND Architects Collective, Wolf Bavaria

KEY DATA

Thermal conductivity: 0.24 W/m²K*
*Previous Darmstadt prototype
'House Made of Paper'

Paper-based materials are currently only used to a limited extent in buildings, and there is a considerable need for technical research in order to develop new products and open up markets. Paper materials are generally used in buildings as ancillary and sub-components, but not as a stand-alone construction material; the work of Shigeru Ban is the notable exception, with his pioneering use of cardboard tube technology, where helically wound tubes of narrow paper strips can be manufactured to different lengths, thicknesses and diameters and with differently performing paper (waterproof, etc). More typically, paper is used as the surface of gypsum plasterboard (drywall), with a paper film on both sides to reinforce the stability of the boards,

1.a Cardboard tube
1.b Flat cardboard profile
1.c Hydro-tee water-resistant solid board for humidity barrier
2.a PhonsStar board for acoustic insulation
3.a Corrugated cardboard layered with solid cardboard
3.b Rectangular cardboard tubes
3.c Solid board and honeycomb composite reinforcement
4.a Corrugated cardboard layered for thermal insulation
5.a Solid board composite for fire insulation
5.b Paper wallpaper

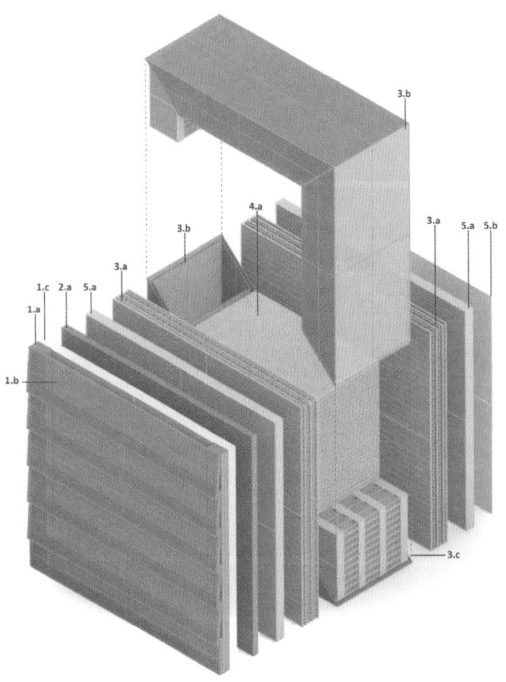

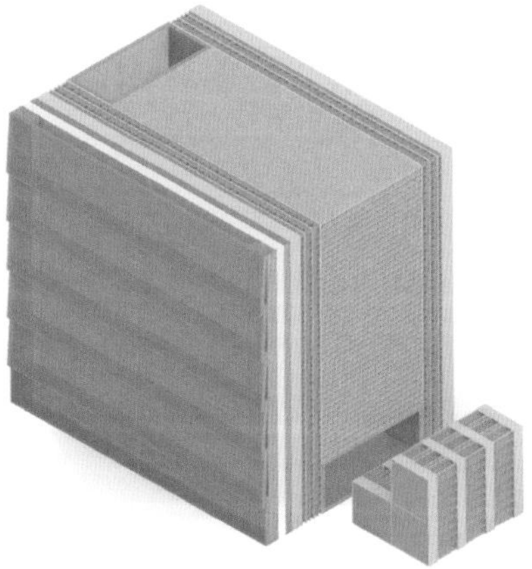

'In addition to constructive feasibility studies, our main focus is on the experimental and numerical analyses of the load-bearing behaviour and structural-physical parameters of components made of paper.'

Dr Nadja Bishara, TU Darmstadt

and as the honeycomb cardboard cores of lightweight timber interior doors. Paper waste is also used as an insulation material, with chopped cellulose used as blown-in thermal insulation. However, the use of paper as a building material for exterior walls is a very new field of application.

This design and research project aims to design, test and develop an external wall system made of paper that will be suitable for use in one- to three-storey buildings. The proposed exterior wall is realized through a structure of functional layers made of paper. In addition, special attention is paid to the development of joining techniques that hold the individual material layers as well as wall components together and simultaneously fulfil mechanical and hygric functions. Of particular importance is the conception and implementation of a fire and moisture protection strategy. The project includes a concept and planning phase, the production of samples at small and full scale, as well as the implementation of physical, structural and mechanical tests. The wall construction is to be developed with the help of prototypes, and an active and passive fire protection concept will be tested. At the end of this research project, ideas for the upscaling of production are to be created and a functional wall system will be available.

Key technical challenges identified by the Darmstadt team include the structural (load-bearing capacity, compressive strength and fatigue strength), fire protection, weather resistance, acoustic properties and thermal insulation.

With the construction of wall elements from different layers it is important that the selection of the layers provides suitable structural and material properties to meet construction requirements. The use of water-repellent components and other specialist coatings is essential to exclude moisture penetration and ensure the necessary level of fire protection. In order to verify the technical performance of this paper-based construction method, the design team also developed a series of test procedures for accelerated ageing of the paper building elements. The team also developed methods for manufacturing the elements (at a laboratory and pilot scale) to explore different ways of connecting the layers of the wall build-up and connecting the wall elements during construction.

With this project the team hope to bring attention to the advantages of using paper as a construction material, including the well-established audited sustainable sourcing, recycling and waste streams.

OPPOSITE Design of the first prototypes of the project, to be built as a 50 x 50cm (19¾ in square) sample of the wall representing all the possible functional layers of cardboard.

3.3 Hemp, Bamboo, Grasses & Palm

'It does create a very soft light and an incredibly soft acoustic, it's as though the room is giving you a great big hug.'

William Stanwix

Hempcrete: hemp lime composite building material

William Stanwix and Alex Sparrow

KEY DATA

Thermal conductivity: Hempcrete wall thickness (250mm/10 in) –0.23 W/m²K, (400mm/15¾ in)–0.15 W/m²K

In *The Hempcrete Book*, experienced hempcrete builders William Stanwix and Alex Sparrow brilliantly explain the technical and environmental benefits of building with hempcrete. Hemp is a fast-growing and deep-rooting plant which is considered good for soil and also acts to naturally suppress weed growth because of its speedy propagation. The UK was once home to a large hemp and flax industry for the production of rope, sailcloth and other fabric. Replaced by imported cotton and jute fibre, the UK hemp industry was all but closed when the cultivation of hemp was banned in the late 1920s because of its potential use as a recreational drug. Since then, types of 'industrial hemp' have been developed which contain negligible amounts of the

BELOW Newly pressed hemp bricks being produced at a factory in South Africa. The world's tallest building constructed from industrial hemp will soon rise in Cape Town.

'Hempcrete is the popular term for a hemp lime composite building material. It is created by wet-mixing the chopped woody stem of the hemp plant (hemp shiv) with a lime-based binder to create a material that can be cast into moulds. This forms a non-loadbearing, sustainable, 'breathable' (vapour permeable) and insulating material that can be used to form walls, floor slabs, ceilings and roof insulation, in both new build and restoration projects.'

William Stanwix and Alex Sparrow

---

psychoactive compound THC, and as well as their use as a construction material they can be used in the production of textiles, paper, bioplastics, health supplements and fuel. Hemp is rich in cellulose, proteins and oils, so one plant can be used for a variety of purposes, with the lowest value hemp shiv (the hard stems of the plant) used in the form of hempcrete for construction. The cultivation of industrial hemp in the UK is now permitted, and was legalized in the US in 2016, and with the necessary licences and renewed interest in its high-value use as textile fibre it will hopefully become a more common crop.

Hempcrete is a mixture of hemp shiv with a hydraulic lime binding agent, and it is typically used as a non-load-bearing mass material wrapped around or infilling a timber frame structure. Hempcrete is usually cast into a lightweight timber formwork, where it is placed by hand into the mould. The hemp shiv, lime and water mixture is stiff and as such cannot be poured, so while hempcrete is considered a carbon-negative material it is labour intensive to build. Hempcrete can also be sprayed into the mould using a compressor and gun.

One of the great benefits of a solid (non-load-bearing) wall of hempcrete is that it has excellent thermal insulative properties with low thermal transmittance as well as providing a breathable building envelope, which is good for the building structure and occupants alike. Hempcrete acts as a hygroscopic material and so it can absorb and re-release moisture, regulating internal humidity and air quality. Hempcrete is also a non-toxic material that does not emit harmful volatile organic compounds (VOCs). Hemp fibre is also used as an insulation quilt and provides similar levels of thermal insulation to that of mineral wool and glass-fibre.

OPPOSITE Hempcrete is also sold as prefabricated blocks, and sometimes a combination of these and the hempcrete mixture as a render is used, with the block wall acting as formwork.

Flat House: hemp & flax fibre house

Practice Architecture and Material Cultures; construction by Oscar Cooper and William Stanwix
Location: Cambridgeshire, UK

KEY DATA

Embodied carbon: –5.1 tonnes (5.6 US tons) with sequestration

Energy performance certification: (A) 104 out of 107 points

Environmental impact rating: (A) 119 out of 120 points

Fire: corrugated panels BS EN 13501-1 to Class C, s2, d0

The Flat House is a beautifully designed low-carbon three-bedroom house prototype that incorporates flax and hemp in its construction. The project was constructed at Margent Farm, a hemp farm in Cambridgeshire, UK, which is committed to organic and regenerative farming methods. The exterior of Flat House is clad in corrugated panels made from hemp fibres and bio-based resin. The panels were developed at the farm and utilize their own harvested hemp as the bulk construction material along with farm waste bio-resin made from corn cob, oat hulls and the sugar waste bagasse.

The panels are fabricated with hemp fibres that are woven into a mat and impregnated with resin before being heat pressed at 180°C (356°F). The panels are used as an exterior rainscreen for the prefabricated hempcrete and timber cassette construction

BELOW Prefabricated timber panels with hempcrete infill are lifted into place; the hempcrete is left exposed internally, which helps to control internal humidity and acoustics.

developed with natural builder William Stanwix (see p.151). The structural panels of the house consist of a timber framework with hempcrete infill which is a mixture of hemp shiv and lime. The hempcrete is left exposed to the interior, which provides an excellent quiet acoustic as well as regulating interior humidity levels. The prefabricated timber panels took ten days to make, with the casting of the hempcrete (with incorporated electrical conduit) an additional four days. The panels were installed in five days and, as Stanwix explains, 'In 20 days of work the structural frame, 35 cubic metres of hempcrete, all electrical first fix and the finished surface were complete, constructed and dry.'

Margent Farm is actively promoting hemp as a sustainable crop which sequesters carbon from the atmosphere at an impressive 15 tonnes (16½ US tons) of CO_2 per hectare. Hemp can be used as a fibre (one of the farm's other collaborations is with the luxury bag company Ally Capellino), the durable hemp stems can be used as the key component of hempcrete building material, and the hemp seeds are edible and rich in healthy omega-3 fatty acids. The cultivation of low tetrahydrocannabinol (THC) hemp crops still requires government licence, and its association with drug use can be counter-productive in the wider acceptance of this valuable, versatile and ecologically sustainable crop.

BELOW Externally the hempcrete walls are covered with specially fabricated corrugated panels made from compressed hemp and sugar waste bagasse.

Bamboo construction

Jan Balbaligo (Carbonmade) and Samsul Aripin (Zewa Architects)

Jan Balbaligo is an architect, researcher and educator who has worked with bamboo across the world. She also created the brilliant 'regenerative cycle' diagram of bamboo (see p.25) which documents the growth of the material, its processing, use, reuse and ultimate end of life recycling. Balbaligo is one of a handful of architects and makers who have become hugely valuable advocates for sustainable and regenerative construction. She describes herself as a 'natural builder' and has lectured widely about bamboo and timber construction. She has also led workshops exploring the technical challenges and limits of building with bamboo, creating permanent buildings and temporary pavilions and event structures. We must not underestimate the value of those like Balbaligo, who combine theory and practice and inspire new generations of designers to think about material and construction choices.

Bamboo has been described as 'green steel', with tensile strength that matches steel and an excellent weight-to-strength ratio. It is not a type of timber but the largest member of the grass family, and one of the world's fastest growing plants with some species growing up to 1200mm (4 ft) per day given the correct growing conditions. The bamboo used for construction typically take 3–4 years to grow and be ready for harvesting as opposed to 25–30 years required for a construction softwood such as pine.

As a robust and aggressively growing plant species, bamboo is easily harvested and quickly regenerates. It grows in every continent in a large belt north and south of the equator, from northern Florida to southern Argentina. A bamboo grove generates 35% more oxygen than an equivalent area of forest, and has been used for phytoremediation in the organic cleaning up of contaminated water and land.

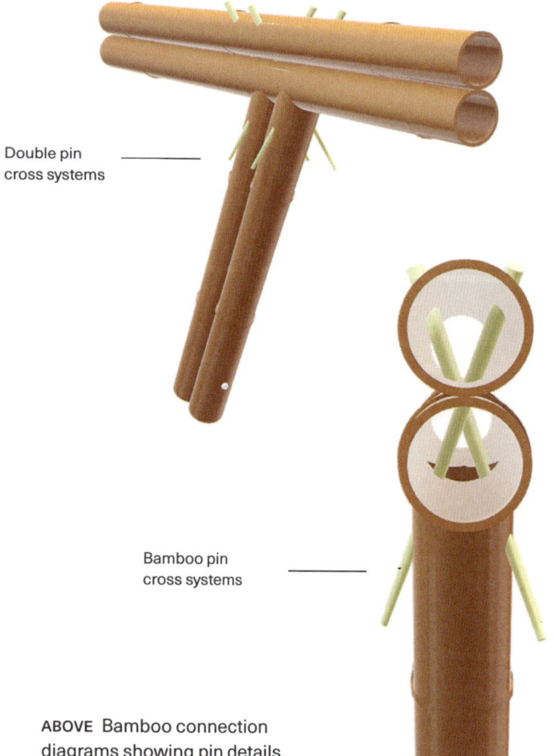

ABOVE Bamboo connection diagrams showing pin details.

OPPOSITE A craftsperson carves bamboo poles using simple tools to create a complex 'fish mouth' connection joint.

'The aim is to regularize and standardize bamboo and make consistent products, and, in the end, architecture and buildings.'

Arief Rabik, Chairman and Founder of Bamboo Village Trust

As a construction material bamboo is lightweight and easy to transport and can be used in temporary and permanent construction projects. It can be bent into curved shapes and bundled to form high-strength composite columns and beams. To prevent against water rot and insects, bamboo does need to be chemically treated; however the preservation treatment is relatively non-toxic in relation to equivalent timber treatments and uses the natural pesticides of boric acid and borax. The bamboo poles are either submerged in baths of a solution of boric acid and borax for seven days and then dried for a further week, or the poles are pressurized with the liquid pumped through the culm (the woody stem). Interestingly, the best time to harvest bamboo is during a new moon, when the bamboo contains less water and is at its lightest. Bamboo does not require complex machine tools for the cutting and joining of components, and a type of modified machete and cast steel splitter are common hand tools used for cutting and forming joints. There are also great examples of more highly engineered connection nodes that are mechanically fixed or bonded with concrete or adhesives in the ends of the bamboo poles.

Bamboo can also be used in a more processed form for flooring, concrete reinforcement, replacing steel and as a fibre in textiles and paper-like products such as tissues. The plant's culm is marked by the distinctive nodes and internal septa that subdivide the hollow interior. As it matures, the walls thicken and become stronger. Guadua and asper are two types of bamboo popular in construction, with diameters of 100–150mm (4–6 in).

BELOW A five-way structural bamboo connection featuring 'fish mouth' connections.

Bamboo & timber composite gridshell dome structure

**Professor Hexin (Johnson) Zhang
(Edinburgh Napier University)
Location: Guangxi, southern China**

In July 2013, engineering and construction students from Edinburgh Napier University (ENU) worked alongside colleagues from Guangxi University of Science and Technology and Inner Mongolia University of Science and Technology to construct a bamboo-timber composite gridshell in Guangxi. This 12 x 12m (40 x 40 ft) square-based dome is unique in its use of a new bamboo and timber composite developed by the team at ENU, and used five layers of the specially fabricated composite bolted together at the nodes to create a three-way structural grid.

BELOW This square-based dome structure is a three-way grid-shell made from a unique lamination of timber and bamboo.

'A good option is to combine the high-strength engineered bamboo and softwood timber to produce a sandwich glulam bamboo-timber beam. The difference in mechanical properties and strength between engineering bamboo and softwood makes them a perfect pair to examine this new concept of bio-based prestress technology.'

Professor Zhang, Edinburgh Napier University

Professor Zhang had observed how prestress technology is used in reinforced concrete to reduce deflection and increase structural stiffness and span-to-depth ratios. In timber structures prestress is also used, typically with the addition of steel tension rods or cables, but Zhang wanted to explore whether an entirely bio-based prestress system could be developed. By using a combination of bamboo and timber elements laminated together, he produced a series of material samples that provided a proof of concept for bio-based prestress. Zhang uses the known technology of glue lamination (glulam) to build up structural components such as prestressed columns or beams. By laminating materials of different strength, density, pre-curved, etc., when they are glued together and then released from a mechanical jig the prestress is activated. This fine-tuning of a bio-based composite means that the mechanical properties can be designed to suit their purposes, and locally introduce stiffness or elasticity around joints and connections, for instance.

Bamboo is comparable, or even higher strength, than the strongest hardwoods but has a lighter self-weight. Bamboo also grows very fast, with only one-sixth to one-tenth of the growing time compared to hardwoods. Zhang thought that an engineered bamboo and softwood timber composite construction would utilize the economy and sustainability of each material while exploiting their different structural properties. The term 'engineered timber' is typically used to describe timber substrates built up of layers and designed for uniform structural properties, and Zhang's innovation demonstrates that a new bio-based composite of bamboo and timber can be engineered with local prestress as well. He hopes that the knowledge and industrial experience obtained from this research will enable engineers of the future to use multi-species bio-based composites for larger-scale constructions such as long-span or tall buildings.

OPPOSITE A student demonstrates the scale of the structural frame, made from laminated timber and bamboo lathes.
LEFT Students collaborate in the construction of the lightweight gridshell canopy structure.

Sugarcrete®: sugarcane waste construction blocks

University of East London

KEY DATA

Dimensions: 100 x 100 x 100mm (4 x 4 x 4 in): varies with application

Embodied carbon: –0.161 CO_2e/kg

Compressive strength: 2.76 N/mm²

Fire performance: 60 minutes minimum

Thermal conductivity: 0.065 W/m²K

Fabrication process: Cast into a mould

The development of Sugarcrete® is a wonderful example of a local collaborative initiative, albeit with a necessary international connection. The University of East London's (UEL) Master of Architecture and Sustainability Research Institute (SRI), with the support of local manufacturer Tate & Lyle Sugars and architectural firm Grimshaw, has developed an innovative low-carbon construction material employing an arable waste product and the clever use of geometry.

Bagasse is the name for waste sugarcane fibres left after sugar sap extraction. Sugarcane is the world's largest crop by production volume, with almost two billion tonnes (2.2 billion US tons) annually produced and subsequently refined as sugar, ethanol fuel or bioplastics. The estimated 600 million tonnes (660 million US tons) of bagasse waste are currently used in part as incinerator fuel or locally burnt as waste management. The ambition of the University-led team was to not only develop a new bio-based material, but to also create a viable construction system.

The bagasse fibres are mixed with sand and other mineral binders and moulded into handleable blocks to minimize material waste and slot together to form wall, floor or roof panels. Inspired by the flat vault system invented in 1699 by French engineer Joseph Abeille, the truncated pyramidal Sugarcrete® blocks reciprocally interlock to form a rigid surface.

LEFT The sugarcane fibres are mixed with binders and pressed into moulds to form easy-to-handle blocks.

OPPOSITE The geometry of the blocks allows them to be easily constructed into an interlocking grid.

'The main innovation with Sugarcrete® is to challenge the common understanding of biomaterials having low structural performance and to develop a system that can be self-supporting.'

Armor Gutiérrez Rivas, Senior Lecturer in Architecture at UEL

This repeatable system is then post-tensioned in small modules with thin steel tie rods, creating impressively strong load-bearing surfaces. No mortar is required for joining the components and the entire system is demountable for relocation and reassembly.

BELOW Members of the design and fabrication team load test a Sugarcrete® floor panel; plan view of the Sugarcrete® panel showing the interlocking pattern.
OPPOSITE Note the steel rods at the edges of the Sugarcrete® floor panel, which are required to post-tension and lock all of the blocks together.

Testing conducted at the UEL SRI laboratories have shown that the Sugarcrete® construction system produces carbon emissions 20 times lower than regular reinforced concrete. Compared with concrete, curing time is one week as opposed to four weeks and it is 20–25% of the weight. Additional analytical testing has shown good compressive strength and fire resistance. The relative low density of the material means that it also performs well as a thermal and acoustic insulator.

The project partners in collaboration with NGOs are currently eyeing potential sites in the sugarcane-producing Global South, where working prototypes can be tested and evaluated.

Hyper-local architecture: building with babassu

German Nieva

German Nieva is an architect and researcher exploring the possibilities of 'hyper-local' architecture, using materials manufactured in part from the non-utilized waste fibres of the babassu palm tree, native to the Cerrado region of midwest, north and northeastern Brazil. Working with a local village community group, he aims to produce prototype structures that show how the local transformation of this waste material can provide materials of excellent strength and durability in a hot and humid climate. Coconuts from the babassu are locally processed to produce flour from the mesocarp layer, and commercially processed to produce lauric oil from its almonds for soap and other high-value cosmetics. The epicarp and endocarp layers are largely disposed of or burnt locally for fuel. Nieva has identified that these waste products are rich in lignocellulose and so can be processed for use in the production of biopolymers for natural composites and plastics.

BELOW Babassu palm tree coconut breakers in the Cerrado region of Brazil on their way to collect the coconuts; the discarded 'waste' babassu coconut shells, which are rich in lignocellulose biomass.

'Localizing and joining up the parts of the construction process have the potential to reduce carbon emissions with the use of locally sourced and manufactured sustainable materials.'

German Nieva

Key to the ambition of the project and the hyper-local approach is not to disrupt or destroy the existing local economy and use of the processed coconuts, but to augment or add new byproducts. Andrew Michler, writing in his book *The Hyperlocalization of Architecture*, explains: 'Good design creates habitat and abundance, is part of a circular economy and improves connections, both human and natural. A less bad design reduces the use of raw materials, it requires less pollution to operate, it creates less harm to the inhabitants.' Nieva wants to use a hyper-local approach to radically reform how we understand contemporary construction. Industrialized building materials such as cement, steel and engineered timber require large facilities with associated energy, labour requirements and carbon emissions. Nieva asks whether a new construction material or suite of materials can be locally produced utilizing the technologies of composites and thermoplastics without destroying the local ecology (natural and economic).

Thus far Nieva has made an informal ethnographical study of the local processing of babassu coconuts in Brazil, and is currently planning for further field studies where materials and processes will be developed in partnership with the local community – a tripartite collaboration between local ecology, technology and user group to research the localized production methods for a new class of biogenic materials.

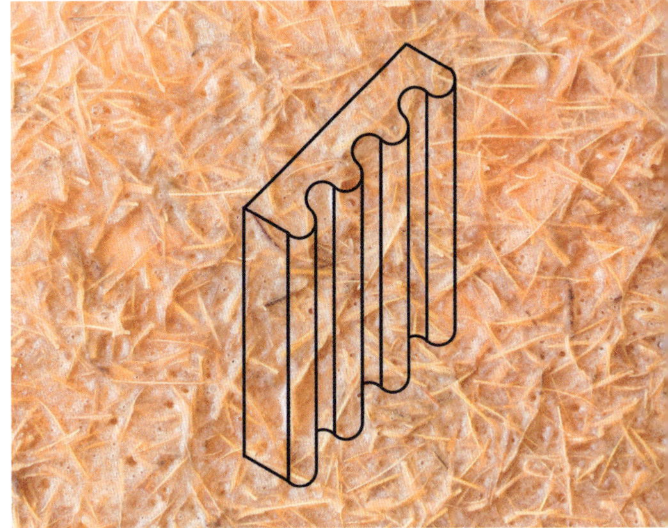

RIGHT Tests of the mechanical properties of babassu composite samples at Bangor University Biocomposites Centre; indicative use of the material as corrugated sheets overlaid with the texture of the babassu fibre-reinforced natural rubber biocomposite.

3.4 Seaweed

'The Modern Seaweed House in Læsø is not only a tale of the renewed use of a remarkable material on a unique site with an extraordinary building history; it is also a crystal ball that catches and illuminates many of the most important issues that face the construction industry today.'

Søren Nielsen, Vandkunsten Architects

The Modern Seaweed House

Vandkunsten Architects
Location: Læsø island, Denmark

KEY DATA

Thermal conductivity: 0.036–0.048 W/m²K (eelgrass has approximately 80% of the insulative value of mineral wool equivalent)

Fabrication process: Compressed into net cylindrical bags

When Danish architects Vandkunsten were asked to design a small summer house on the island of Læsø they looked no further than the neighbouring building for inspiration. Kaline's House is a 150-year-old historically preserved longhouse that had recently been restored and is a rare example of the local seaweed thatch. The local vernacular sees a hugely thick layer of seaweed piled on top of the house, providing both a waterproof and thermal hat.

The Modern Seaweed House was built in part as a construction experiment and not, as the designers make clear, a social experiment. To that end, the form of the structure is typical local roof pitch size and orientated east–west. The timber-frame construction is insulated with seaweed replacing mineral wool, and is then overclad with additional pillows of seaweed which is packed into cylindrical bags and held in place by woollen nets. Different densities of material are used for roof and walls, with the roof pillows

BELOW Eelgrass is packed into woollen nets to be used as external insulation; prefabricated timber panels are filled with eelgrass insulation.

thicker and less dense and the wall panels smaller, more tightly packed and sitting within prefabricated timber panels.

The 'seaweed' used is in fact eelgrass, not a seaweed at all but a plant that grows in shallow water. Eelgrass meadows provide important marine habitats and have become increasingly endangered. The eelgrass used in Læsø, however, is not farmed but is collected from beaches as an abundant locally available waste material. Vandkunsten partner Søren Nielsen explains that 'Eelgrass has an insulation value of 80% of mineral wool when optimally

compressed… It is non-toxic. The smell is very pleasant, like fresh hay. It does not attract pests, besides some birds who like to nest in it. It can be burnt passively, but it cannot flame and does not nurture fire. The acoustics are great and make it suitable as a noise-absorbing agent.'

Studio Two of the world-famous Abbey Road recording studios in London used to feature acoustic hanging panels of Cabot's Quilt, a fabric or paper quilt filled with eelgrass. Examples of eelgrass thermal insulation can be found up and down the east coast of North America, notably in Halifax, Nova Scotia. Eelgrass is a hygroscopic material which can absorb and emit moisture, helping to regulate internal building humidity. While Vankunsten Architects' one-off contemporary showcase for eelgrass is an interesting 'for instance', both re-imagining vernacular construction and using local waste material send an important message.

BELOW Eelgrass cylindrical bales cover the roof and walls of the house; cross section of the external cladding.

OPPOSITE The Modern Seaweed House on Læsø island has become an inspiration for a new generation of environmentally conscious designers.

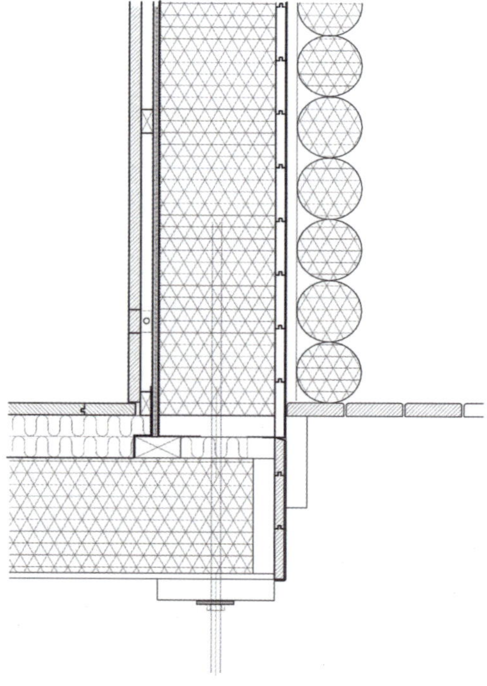

3.5 Mycelium

'Innovative natural materials, grown by means of microbial fermentation, hold the promise for the creation of a near future where human activities and the rhythms of the larger ecosystem are not in conflict with each other.'

SQIM

Mycelium construction materials

SQIM and Mogu

KEY DATA

Dimensions: 100mm (4 in) thick

Density: 100 kg/m³

Embodied carbon: Carbon negative

Compressive strength: 0.01 N/mm²

Fire performance: B-s2-d0 /D-s2-d0
(Fire Reaction UNI EN 13501-2)

Thermal conductivity: 0.024 W/mK

Fabrication process: Grown into
a mould

Mycelium is the vegetative root system of a fungus, with mushrooms being the fruit. As a construction product, mycelium is not a material in and of itself and needs calorific content (food) of another substrate to grow. What is so incredible about mycelium is that it is not discerning about the types of foodstuffs and so waste products from agriculture and other industries (including timber and paper) can provide excellent substrates on which to feed and bind together. Product design firm Mogu, in collaboration with technology developer SQIM, produces materials by growing mycelium on low-value agricultural industry substrates like cotton and hemp waste and transforming them into a durable high-value construction product. This symbiotic relationship with a local waste stream is key to the development of a construction industry circular economy.

BELOW Mycelium is put into a mechanical heat press to form a hard, dense substrate for flooring.

To produce mycelium, waste material is initially sterilized to prevent contamination and the mycelium powder is added. The material is then left to grow in plastic bags at room temperature. When the mycelium has taken hold, this material is ground up and placed in reusable moulds. Once the mycelium has bound all the waste substrate and filled the mould, the mycelium components are then slowly dried to create a lightweight inert material. Heat of no more than domestic oven temperatures can also be used for this last process, but this unnecessarily adds to the embodied energy. Mycelium is fast growing and carbon neutral (or negative) depending on the U-value substrate, and it has useful material properties such as low thermal conductivity, good acoustic absorption and lightness, with a similar mass to that of expanded foam but non-toxic and made without the use of petrochemical binders or propellants, and thus VOC-free.

Mycelium is not dense and as such is not a natural load-bearing or compressive material. However, Mogu has developed a mycelium flooring material where, rather than being moulded, the substrate bound with mycelium is compressed with heat (180°C/356°F) to form a hardwearing and non-toxic alternative to vinyl and rubber floor coverings.

BELOW Mycelium is incubated with waste substrate (feedstuff) in plastic bags to retain moisture; mycelium is heated and dried to create an inert construction material.

OPPOSITE Mycelium is grown into vacuum-formed moulds; samples of heat-treated mycelium for use as flooring and board products.

Tree Column: 3D-printed mycelium

**Paola Garnousset, Martin Detoeuf and
Pierre de Pingo (Blast Studio)**

The Biological Laboratory of Architecture & Sensitive Technology (Blast Studio) was established to explore the relationship between nature and technology and demonstrate how waste material can be usefully transformed into architecture. Blast has utilized the 3D-printing technology more typically employed in construction to extrude clay or concrete to print a paper pulp slurry made from discarded coffee cups which is seeded with mycelium (fungal root) spores. This unique mycelium fabrication process obviates the need for any formwork, with mycelium typically cultivated and 'grown' within a rigid mould.

After the paper pulp mycelium mix has been printed, the mycelium grows through and around the surface, thickening, binding and stiffening the structure. Blast has also cultivated a version of these structures which, when left, will begin to 'fruit' with edible mushrooms growing from the surface. In order to solidify and stabilize the mycelium, the structure

BELOW The forms of the printed pieces were inspired by tree trunks, and the crevices help to shield organisms; printing a mycelium column from a paper pulp slurry seeded with mycelium spores.

'In nature nothing goes to waste, the scraps of an organism become the resource of another one. This circularity is at the heart of our work: cities' waste is our resource to create the interiors of tomorrow.'

Blast Studio

needs to be heated or dehydrated, which kills the mycelium organism and leaves a benign substrate.

As part of the *Waste Age* exhibition at London's Design Museum in 2021, Blast displayed their mycelium Tree Column, which was over 2m (6½ ft) tall. The profile and cross-sectional design were developed with algorithmic software, which created a tree-like form that flares out at its base for stability and has a dense undulating cross-sectional profile – again creating stability, but also the perfect incubation environment for mycelium growth. The tower was 3D printed in sections which are then fused with the aid of the mycelium root network.

Blast Studio continues to develop their use of waste materials and 3D printing with mycelium is being used as a way to process and homogenate the material structure. Blast is currently developing a larger-scale pavilion structure to show how waste materials can be successfully transformed on an architectural scale. The Root Pavilion will comprise three tree-like columns which combine into an arched dome, with the structure composed of waste cardboard, crushed brick dust and sawdust. Blast Studio explain: '...with this pavilion, we can create an architecture for all living organisms. Bees and other pollinators inhabit the multiple dark interstices of the pillars while plants grow over the roof and fungi can inhabit some of the humid crevices of the columns where they sprout. From local waste we can build an architecture that creates a new ecosystem for the living.'

ABOVE The finished Tree Column, a 3D-printed mycelium sculpture made from used coffee cups.

MycoTree

Sustainable Construction, Karlsruhe Institute of Technology; Block Research Group, ETH Zürich; Alternative Construction Materials, Future Cities Laboratory, Singapore-ETH Centre; and MycoTech

KEY DATA

Material density: 440 kg/m³

Compressive strength: 0.61 N/mm²

BELOW Sawdust, woodchip and food industry waste is mixed with mycelium spores to form a spongy mass. This mixture is then placed in moulds to densify.

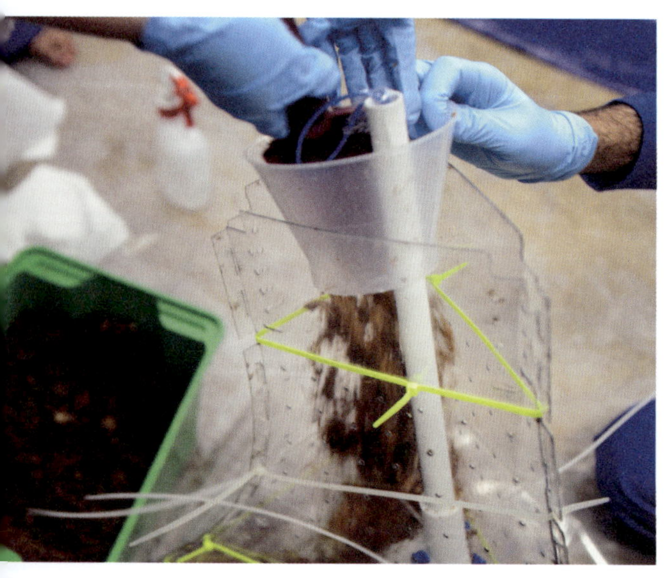

MycoTree is the result of a collaboration between teams led by Dirk Hebel at Karlsruhe Institute of Technology (KIT), Germany, and the Block Research Group at the Swiss Federal Institute of Technology (ETH) Zürich led by Philippe Block. They were invited to design a structure for the inaugural Seoul Biennale of Architecture and Urbanism 2017 in South Korea. This experimental structure was designed and built with students to demonstrate how grown and cultivated materials can challenge the orthodoxy of the mined and processed materials typical of the contemporary construction industry.

This tree-like branching structure is made from interconnected mycelium elements which support a 4 x 4m (13 x 13 ft) bamboo grid at a height of 3m (10 ft). The bamboo grid weighs 134kg (295 lbs), and the overall weight of mycelium components amounts to 182kg (400 lbs). The form-finding of the structural geometry was informed by several key design constraints, such as limiting the length of a single mycelium member to 600mm (23½ in) to avoid buckling, with all structural nodes of no more than four connection points. The geometry of the structure was designed using 3D graphic statics, a novel method developed by the Block Research Group at ETH Zürich that extends the traditional 2D structural design technique, allowing for the exploration of efficient and expressive structural forms beyond the compressive arch or vault.

The moulds were made from interlocking laser-cut acrylic sheets and the mycelium is then grown within them. The mycelium was locally cultivated on wood chip, sawdust and food industry waste products such as sugarcane and cassava root. This waste substrate is mixed with mushroom spores and after a few days starts to form a slightly spongy mass. The mycelium mixture is placed into the moulds where it continues to densify, eventually forming a tough protective outer skin. The mycelium components are then removed from the moulds and dehydrated, which kills the organism and thus stops the growth process.

'It takes two weeks to grow any form, any shape you would like …
It only needs a form, a little bit of biological residue and a little bit of
knowhow, and then you can grow this in any place you can think of.'

Dirk Hebel, Karlsruhe Institute of Technology

To create precise engineered connections and load transfers from one mycelium element to another, the ends of each mycelium component are capped with thin sheets of a bamboo composite material. These sheets are also used for compacting the mycelium mixture into the mould and are eventually fixed to the mycelium components with dowels. In engineering terms mycelium is not a strong material as regards compressive and tensile loading. However, it is a great bulk material that can be cheaply (and locally) produced with waste products from agriculture and other industries such as paper and packaging. The MycoTree project demonstrates how innovative design and engineering add value and utility to these sustainable building products of the future.

BELOW The mycelium branching structure supports a light bamboo framework.

3.6 Algae

'By working with nature to use existing microalgae to bind minerals and other materials together to create new types of sustainable biocomposite building materials, we can eliminate most, if not all, of the carbon emissions associated with traditional concrete-based building materials.'

Wil V. Srubar III, University of Colorado

Bio-Block Spiral: bio-cement blocks from microalgae

Prometheus Materials & SOM

KEY DATA

Embodied carbon: Net zero

Compressive strength: 16.55–24.13 N/mm²

Prometheus Materials was born out of a research project that started in 2016 at the University of Colorado, Boulder. In response to a US government initiative relating to engineering living materials, the team wanted to utilize microalgae as a building material. Prometheus has developed a technology that harnesses microalgae as a replacement for carbon-intensive Portland cement, and uses a variety of biomineralizing microalgae to make their products. These microorganisms are non-toxic and are grown using sunlight, water and CO_2. Prometheus describes the patent-pending process as 'photosynthetic biocementation', and says it is similar to the natural formation process of coral and seashells. A natural photosynthetic process is used to produce the algae, so instead of producing CO_2, CO_2 is sequestered during the manufacturing process.

BELOW Microalgae produced by Prometheus via a natural photosynthetic process which sequesters carbon; construction blocks are mechanically compressed, made from sand aggregate and water with microalgae to replace the Portland cement.

'Coral reefs, seashells and even the limestone we use to produce cement today show us that nature has already figured out how to bind minerals together in a strong, clever and efficient way.'

Wil V. Srubar III, University of Colorado

Prometheus has used this cement-replacement material in the manufacture of concrete blocks, using the same production process as standard blocks, with sand, aggregate and water, but the bio-cement replaces Portland cement. The benefit of this approach is the ability to use construction industry standard infrastructure and technology to fabricate, transport and construct. By replacing regular blocks like for like, in this instance, material science and sustainable construction do not look any different to standard building practice. While this approach might not capture the imagination in the same way as some of the other more novel featured materials and designs, the impact of this kind of innovation is potentially massive.

Prometheus Materials has subsequently partnered with multi-disciplinary architecture studio Skidmore, Owings & Merrill (SOM) to develop Bio-Blocks; in 2023 they collaborated to construct the Bio-Block Spiral for the Chicago Architecture Biennial in order to showcase the utility and durability of these 'cement-free' building blocks. As Prometheus relies on natural, photosynthetic processes to manufacture its products, they can be produced with net-zero embodied carbon emissions (per Life Cycle Assessment stage A1–A3/cradle to factory gate[1]) inclusive of biogenic carbon.

RIGHT Bio-Blocks produced by Prometheus in collaboration with SOM have a satisfying algae-green hue.

Air Bubble: air-purifying eco-machine

ecoLogicStudio

KEY DATA

Embodied carbon: Net zero

Compressive strength: 16.55–24.13 N/mm²

Architects and educators Claudia Pasquero and Marco Poletto established ecoLogicStudio in 2005. The studio sets out to explore the relationship between biotechnology and architecture, and tests how microorganisms such as algae can be incorporated into the building fabric to condition the internal environment.

As part of the UN Climate Change Conference in Glasgow (COP26) ecoLogicStudio was commissioned to create a pavilion that would demonstrably incorporate algae as part of an air-purifying system. Air Bubble, an air-purifying eco-machine, is a small pavilion or air structure made from 23 conjoined air-inflated ribs. Over 10m (33 ft) in length, the low-pressure inflatable is anchored to the ground

BELOW Isometric rendering of the Air Bubble air-purifying eco-machine.

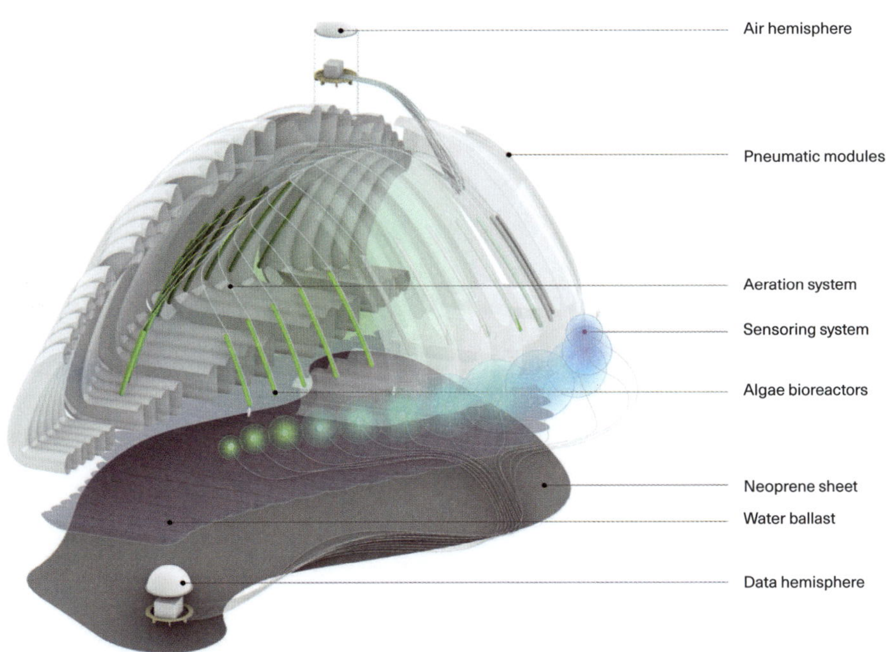

by water ballast and incorporates 24 algae-filled bioreactors (12 on each side) integrated into the transparent polymer skin. An algae photobioreactor is a device that utilizes carbon dioxide, water and algae to produce biomass and oxygen and is powered by the ultraviolet light of the sun.

To activate the air-purifying capability, air is pumped or bubbled through the photobioreactor, which absorbs six key air pollutants including particulates PM2.5 and PM10, nitrogen dioxide (NO_2), sulphur dioxide (SO_2) and carbon monoxide (CO). The designers explain that the pavilion is capable of absorbing 97% of the nitrogen and 75% of the particulate matter in the air.

The Air Bubble is ecoLogicStudio's first inflatable bioreactor. It contains 6000 litres (1585 gallons) of water for foundation ballast and 200 litres (53 gallons) of Chlorella vulgaris, which is a green fresh-water alga capable of filtering 100 litres (26 gallons) of polluted urban air every minute. The lightweight structure is made from pattern-cut TPU membrane welded together to form a mollusc-like cellular form. While the membrane is a petrochemically based thermoplastic polyurethane, its overall weight of less than 100kg (220 lbs) is tiny compared to an enclosure of comparable size and of conventional construction. There are currently similar polymer membranes with up to 70% plant-oil resin content, and the reuse of plastic membranes to create low-pressure inflatables has been successfully explored by Argentine artist Tomás Saraceno with his Museo Aero Solar, a flying sculpture made from plastic bags, for the 2021 Venice Biennale.

BELOW Interior view of Air Bubble with the algae cells clearly visible on each side of the structure.
OPPOSITE The inflatable Air Bubble structure formed by air-inflated cells is held on the ground with water-filled foundations.

3.7 Biopolymers & Recycled Polymers

'Bioplastics, also known as organic plastics, are plastics derived by sourcing monomers from a renewable biomass source like trees, crab shells or potatoes.'

Billie Faircloth, Architect and author of *Plastics Now*

Recycled plastic waste

Professor Paulo Gomes, Federal University of Bahia (UFBA), Brazil
Location: Pedra Furada, Brazil

The Pedra Furada Transformation Workshop is a pilot social enterprise that transforms polypropylene plastic waste into bar and sheet material from which furniture is fabricated and locally sold. It was set up to provide employment for local women. Pedra Furada in Northern Brazil was originally established as a Quilombo, an encampment of formerly enslaved people, and remains one of the poorest villages in the country. The village is surrounded by mangrove forests, which are sadly being destroyed and replaced by aquaculture ponds. While these ponds and associated pesticides are endangering the natural habitat, they do provide employment and an alternative to the brutal subsistence farming of aratu crabs. The Transformation Workshop aims to transform both plastic waste and lives through training and employment, and meets 10 of the UN Sustainable Development Goals[2], combining environmental conservation with social justice.

BELOW Waste plastic is roughly chopped and placed in specially modified steel pans devised by Professor Gomes; this cheap waste plastic is converted into a viable recycled construction material in board and rod form, which can be cut and machined with regular hand- and power tools.

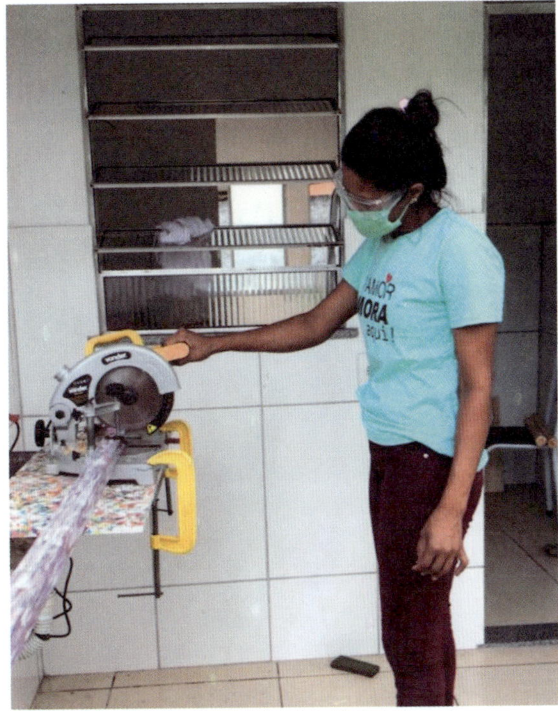

'Plastic pollution is a problem that the current recycling model, based on large industrial plants, has failed to solve. An alternative to this model is so-called local recycling, where discarded plastic is transformed into objects of value in the disposal environment itself, by people without specialized training.'[3]

Professor Paulo Gomes, Federal University of Bahia

The project was developed by Professor Paulo Gomes from the Federal University of Bahia. The machinery he developed to produce the recycled plastic is inexpensive to build, simple to operate, safe and easy to maintain. Gomes has made the plans of the machines open source, and has planned two more Transformation Workshops in Bahia's capital, Salvador.

The waste plastic used for this project is thermoplastic polypropylene (PP), typically used to produce the ubiquitous Monobloc chair. While this plastic is recyclable, it does not have a great local value compared with polyethylene terephthalate (PET) and polyvinylchloride (PVC). Waste plastic recycling rates in Brazil are very low, and as the fourth largest producer of plastic waste in the world only an estimated 1.28% is recycled.[4] Gomes' technical innovation was to use simple, readily available components which include non-stick paella pans and stainless steel tubes. In conjunction with small heating elements, a thermal shield and the use of pressure, ground-up PP can be re-formed into 600mm (23½ in) diameter flat plates or lengths of rod 1m (3½ ft) long. At a working temperature of 180°C (356°F), a medium to high domestic oven temperature, the process is safe and reliable. Gomes is currently investigating the use of organic fibres such as coconut fibre (coir) as an additional ingredient for added tensile strength.

BELOW A child's school desk created from the recycled waste plastic.

OPPOSITE, ABOVE Professor Gomes continues to develop new low-cost tools to produce usable plastic substrate from waste, in this case square panels.

RIGHT A partner in the Pedra Furada Transformation Workshop shows off some newly produced material.

STRONG
WOMAN

Shellmet: shellfish bioplastic

**Quantum, TBWA\HAKUHODO
and Koushi Chemical Industry Co.**

Sarufutsu in Hokkaido is the northernmost village in Japan and the largest domestic producer of scallops, which are locally shelled, packed and shipped. This process can generate up to 40,000 tonnes (44,000 US tons) of scallop shells a year, and whereas these shells were previously exported for alternative reuse, regulatory changes in 2021 meant that a huge stockpile of shells was building up. Left unattended in the ground, these waste shells can cause soil contamination, and new uses for this waste stream were sought. This has led to the development of Hotamet, or the 'Shellmet'.

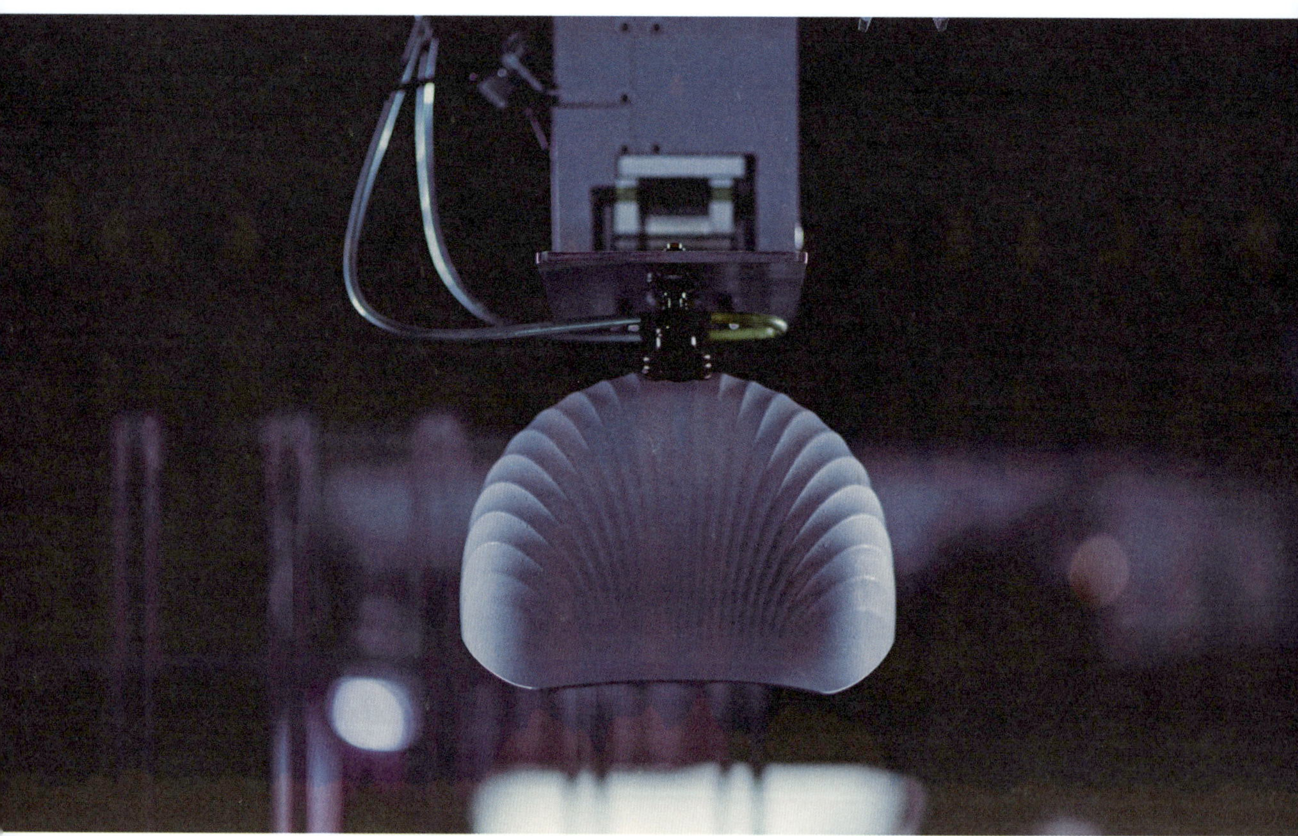

BELOW The manufacturing process of the Shellmet. The geometry and ridged surface of the Shellmet gives it its strength and stiffness.

The Shellmet is a safety helmet made from waste scallop shells. Scallop shells are a form of calcium carbonate, and are much less carbon intensive than other forms of limestone-derived calcium carbonate that can be used for biopolymers. The waste shells are boiled, sterilized, crushed into a powder and formed into pellets with the addition of recycled polymers. The pellets are then used in an injection-moulding process to form the safety helmets.

Using principles of biomimicry, where engineering solutions are inspired by natural processes, the Shellmet utilizes a ribbed surface structure, increasing the structural strength of the protective helmet by 33% compared to a smooth surface. This undulating 'shell-like' texture not only alludes to its underwater origins but also creates strength with minimal material use and thus relative lightness.

Upcycling is an important part in any successful management of valuable waste streams, and innovative projects like the Shellmet demonstrate how high-value and useful products can be successfully manufactured from the 'waste' of other industries. While helping to support the village, this new industry has also averted a potential environmental pollution problem, preventing soil contamination and mitigating shipping and disposal costs (both environmental and financial).

The Shellmet was originally conceived as a protective hat for the marine industry, but there are now plans to extend its application as a standard protection item for use by first responders, cyclists and the construction industry. In 2025 the Shellmet was introduced as an official safety helmet for the World Expo 2025 taking place in Osaka, Kansai, Japan.

'Protect heads, and the earth.'

Quantum start-up studio

Bacterial biopolymers

Dr Pooja Basnett, University of Westminster

LEFT Dr Pooja Basnett holding a sample of biopolymer in her laboratory at the University of Westminster.

The production of petrochemically derived synthetic polymers (plastics) requires huge amounts of energy and associated carbon emissions, and while many plastics are recyclable, they are unfortunately not routinely recycled and single-use plastic products have been widely acknowledged as an environmental problem. When plastics are not fed back into the waste stream and are discarded as waste, they will eventually break down in a marine environment to form microplastics (small plastic grains) that can be ingested by fish, humans and other animals. The deleterious health effects of microplastics are only beginning to be understood, but tissue damage and weakened immune systems have been identified as possible symptoms. Better regulation around single-use plastics and improved recycling rates can certainly help and somewhat limit the use and spread of microplastics, but how can we prevent them in the first place?

Dr Pooja Basnett is a member of the Sustainable Biotechnology Research Group at the University of Westminster, London. The group is developing polymers or plastics that have properties very similar to synthetic plastics, and in fact their properties are almost identical to polypropylene, but they are produced by bacteria using renewable carbon feedstock. These types of biopolymers are known as polyhydroxyalkanoates (PHAs) and are a naturally occurring compound synthesized by a variety of bacteria which is produced via a fermentation process.

PHAs can be broadly classified into two types: short-chain and medium-chain length depending on the number of carbon atoms. Short-chain PHAs are brittle with a high melting temperature, whereas medium-chain length PHAs have good elastomeric properties and have a low melting point. Importantly, PHAs are biodegradable in nature with several

'We are one of the many research groups that focuses on building sustainable, environmentally friendly alternatives to replace synthetic plastics.'

Dr Pooja Basnett, University of Westminster

---

microorganisms that break down (depolymerase) polymers with enzymes – unlike synthetic polymers such as plastic water bottles, which cannot be successfully degraded in nature, with very few bacterial strains identified that can fully degrade synthetic plastics. PHAs are also biocompatible, which means that they are non-toxic to humans.

The fermentation process for the production of PHAs is very similar to the brewing of beer, with a bacterial broth fermented in a bioreactor and fed with carbohydrates. The PHA is then extracted and analysed for mechanical properties such as tensile strength and melting temperature. This base material is then turned into a polymer with downstream processing similar to that of synthetic polymers. PHAs are a range of plastics with different properties, and they have been successfully used as biodegradable sutures in medicine, and Basnett has successfully 3D-printed PHAs. At present the production of PHAs is expensive and slow compared with synthetic polymer production, because of lack of infrastructure; the ambition is to create an economically and environmentally sustainable production process of bioplastics from bacteria that can be more widely used for packaging or construction materials. Using cheaper carbohydrate feedstock is one way of reducing production costs, which also utilizes an untapped waste stream.

BELOW Bacterial biopolymers being produced via the fermentation of rice and vinegar (left) and sushi rice (right). Samples grown by University of Westminster graduate Ross Wilson.

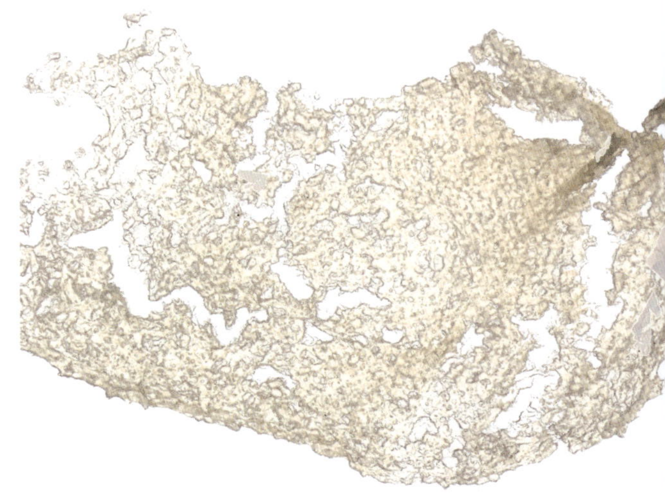

3.8 Wool

'High in water and nitrogen, wool is naturally fire resistant. Wool does not melt, drip or stick to the skin causing burns, and when subject to extreme heat it produces less smoke and less noxious fumes.'

British Wool

Sheep's wool: insulation & composite

Solidwool

KEY DATA

Thermal conductivity: 0.035–0.040 W/mK

Ignition point: >520°C

Reaction to fire: EN13501

Flammability: ISO19925-2

In relation to other commonly used fibres such as cotton, polyester, acrylic and nylon, wool is more flame resistant and is used in protective clothing for fire fighters and the military. Wool has a high ignition temperature and does not melt or release toxic fumes when exposed to high temperatures. Wool is also an excellent natural thermal and acoustic insulation material. Designed to protect sheep in all weathers, wool traps airs within the fibres to provide superb thermal resistance (or insulation). Mountain breeds contain coarse kemp fibre in their fleece, which is particularly good at trapping air within the fibres, improving the insulation properties. Wool is a hygroscopic fibre which can absorb and release moisture, helping to regulate internal temperature when used as part of a 'breathable' building skin.

BELOW Herd of Welsh Mountain sheep, one of the hardy breeds that Solidwool uses as a source of fibre for its sheet material. The breed and wool colour determine the composite colour.

It is a safe material that does not cause allergies or bacterial growth, which is good for both building inhabitants and the builders installing the sheep's wool insulation. Sheep's wool does not emit harmful volatile organic compounds (VOCs), and the UK trade body British Wool explains that wool can actually absorb and trap VOCs in its fibres. Wool is, however, prone to moth infestation; as such it does need to be treated, typically with borax, or with Permethrin, which is toxic to aquatic life. To avoid the use of insecticides, the Austrian company Isolena has developed a plasma-ion treatment which permanently alters the molecular protein structure of wool fibre and prevents the wool being a nutritional source of protein for moths and beetles.

As well as a natural insulation material, wool can also be used as the fibre reinforcement for a bio-based composite material. UK-based company Solidwool has pioneered the use of sheep's wool to manufacture chairs and sheet material, specifically using the wool of Welsh Mountain and Herdwick sheep (hardy native UK breeds); the wool is coarse and strong, but there is currently very little demand for the use of this type of wool in the garment industry. Solidwool chairs are made with a wool-composite seat, recycled steel frame and locally sourced ash timber legs. The equivalent of one whole fleece goes into each chair, which is fabricated using

ABOVE Solidwool produces sheet material in Welsh Mountain (left) and Herdwick (right) wool, which can withstand production processes including milling, drilling, laser cutting and sanding.

OPPOSITE, ABOVE The wool from mountain breeds is often discarded as being too coarse and difficult to dye to be utilized by the fabric and fashion industry.
LEFT Solidwool's Hembury Chair, made of a fibreglass-like composite created from bio-resin and sheep fleece.

composite technique on a mould with a resin binder. The resins that Solidwool are using have a 50% bio-based renewable content and are sourced from the waste streams of other industrial processes, such as wood pulp and bio-fuel production, and are not from food sources or displacing food-based agriculture. The Solidwool chair is an exercise in local sourcing, with the different colours of the chairs directly related to the sheep breed.

3.9 Salt

'It was one thing to create a panel out of salt that works, but it is another whole research project in how you can scale it up and create a farm where you can grow materials rather than produce them in a traditional way… growing objects rather than manufacturing them.'

Atelier LUMA

Wall of salt and salt door handles

**Karlijn Sibbel and Henna Burney
(Atelier LUMA)
Location: Camargue region, France**

Luma Arles was established in 2013 at the Parc des Ateliers, a former railway wasteland covering an area of 27 acres on the outskirts of Arles. Luma Arles is conceived as a new kind of arts centre where visiting artists will participate in interdisciplinary collaborations with engineers, scientists and other specialists, with particular emphasis on the relationship between art, nature and human health. At the centre of the Luma Arles is The Tower, designed by Frank Gehry, which is surrounded by former industrial buildings. Atelier LUMA was established to run the design research programme for the project, and is now located in Le Magasin Électrique, which has been creatively renovated with innovative locally sourced bio-based materials in a collaboration between BC Architects (see Gent Waste Brick, p.56) and Assemble.

BELOW The colour and texture of the salt panels is determined by the weather conditions and water flow of the growing cycle; designers worked closely with the salt workers of the Salins de Giraud, drawing upon their vast experience.

130
120
110
100
90

As part of Gehry's central tower, Atelier LUMA developed internal cladding panels made of salt crystals. Working with local salt farmers, or saliniers, salt ponds are utilized to create building components. Atelier LUMA wanted to use this abundant local mineral and its skilled producers to create architectural components that are grown in the salt ponds. The designers had to be careful how they used this material, which forms a hard crystalline surface, because it can be hugely corrosive to certain materials. They also had to devise a method of fabrication where the salt growth could be controlled to form a consistent panel form. The salt panels were ultimately grown on lightweight perforated metal armatures, which form the backing of the cladding and provide suspension points. For the cultivation process, the panel frames were placed into a larger framework which could be mechanically lowered into the salt ponds. It took 15 days for the salt to form a thick enough layer, with the panels raised out of the salt ponds and rotated

OPPOSITE The production of the salt involves seawater from the Camargue being pumped into giant ponds, and uses natural evaporation to bring seawater to its saturation point.

every day during this period. Henna Burney, material designer at Atelier LUMA, has described this as a 'natural additive manufacturing process'.

The salt panels are fire resistant and have high thermal mass, and they also provide good air quality, acting as a natural antibacterial, antimicrobial and antiviral. Following the successful installation of the panels and years of material research, Atelier LUMA has subsequently developed prototype replaceable salt door handles that utilize the natural antibacterial properties of salt.

LEFT The individual salt wall panels can be removed and restored when necessary, by placing them back into the water of the crystallization plant.
BELOW The 4000 salt panels cover a surface area of 560m² (6028 sq ft) over ten floors of the Frank Gehry Tower.

Endnotes

INTRODUCTION

1. Adriaan Beukers and Ed van Hinte, *Designing Lightness*. Rotterdam: nai010 publishers, 2020

2. Prof Sir Brian Hoskins (Grantham Research Institute on Climate Change and the Environment), BBC Radio 4, *Today* programme, interview with Martha Kearney 14.07.2023

CHAPTER ONE

1. Robbin, T., 1996, p1

2. Heinzelmann, Heinzelmann and Lorenz, 2018, p1

3. Spray-On Fabric: Manel Torres at TEDx, Imperial College. www.youtube.com/watch?v=EW6Gv-loErw

4. Transparency Market Research: Construction Waste Market. bit.ly/4cSUiJh

5. Department for Energy Security & Net Zero, Net Zero Government Initiative: UK Roadmap to Net Zero Government Emissions, UK, December 2023

6. UK Research and Innovation, Industrial Decarbonisation Challenge. bit.ly/3zzM1eC

7. The Construction Material Pyramid www.materialepyramiden.dk

CHAPTER TWO

1. UN Environment Programme, wedocs.unep.org/handle/20.500.11822/44901

2. Nzinga Mboup, lecture at the University of Westminster, School of Architecture and Cities, 21 October 2021

3. Nzinga Mboup, 'Senegal architects ditch concrete for earth in revival of old techniques', Reuters. 21 May 2021. reut.rs/3xDndlz

4. Johanna Lehne and Felix Preston, *Making Concrete Change,* Chatham House Report, 2018. bit.ly/3VNpJxB

5. Embodied carbon of the Gent Waste Brick has been calculated for a 60-year life cycle stages A–D to EN15804+A2:2019. Belgian clay brick data is sourced from TOTEM: EcoInvent Database and is for a 60-year life cycle stages A–D to EN15804+A2:2019. A Gent Waste Brick has 0.17kg CO_2e/kg compared to a Belgian clay brick, which has 0.54kg CO_2e/kg, over a 60-year life cycle

6. Sian Ricketts and Neil Michels, 'In practice: How Carmody Groarke developed a low-carbon brick', *Architects' Journal*, 21 December 2022, bit.ly/3W6ODto

7. Ibid

8. Bini, D., 'A Pneumatic Technique for the Construction of Thin Shells' pp.52–55 in Feder, D. (Ed.), International Association for Shell Structures (IASS), Proceedings from the 1st IASS International Colloquium on Pneumatic Structures. Stuttgart, May 11th–12th 1967. Stuttgart: IASS, 1967, p.52

9. Tony Whitehead, 'Innovation: Biozeroc', *Concrete Quarterly*, Winter 2022, Issue Number 281

10. Graeme DeBrincat and Eva Babic, *Re-thinking the Lifecyle of Architectural Glass,* Arup. bit.ly/4ckwr5c

11. 'Sand: Rarer than One Thinks', UNEP Global Environmental Alert Service (GEAS) March 2014. bit.ly/4eScPqD

CHAPTER THREE

1. Life Cycle Stages. bit.ly/3xMrgMj

2. United Nations Department of Economic and Social Affairs, Sustainable Development Goals. sdgs.un.org/goals

3. Adriano Puglia Lima, Rosana Lopes Lima Fialho and Paulo Alberto Paes Gomes, 'Social Technology for Local Recycling of Plastic: A Model of Circular Economy and Sustainable Development'. VIII International Symposium on Innovation and Technology, Circular Chemistry and Circular Economy, 2022. ISSN: 2357-7592

4. Kaza, Silpa et al. 'What a waste 2.0: a global snapshot of solid waste management to 2050'. World Bank Publications, 2018

Bibliography

BOOKS AND PAPERS

Arup + Material Cultures, *Circular Biobased Construction in the Northeast and Yorkshire*. Arup and Material Cultures, 2021

Ashby, M.F., *Materials and the Environment: Eco-informed Material Choice (3rd edition)*. Oxford: Butterworth-Heineman, 2020

Beukers, A. and Ed van Hinte, *Designing Lightness: Structures for Saving Energy*. Amsterdam: nai010 Publishers, 2020

Hebel, D., Marta H. Wisniewska and Felix Heisel, *Building from Waste: Recovered Materials in Architecture and Construction*. Basel: Birkhäuser, 2014

Heinzelmann, D., Heinzelmann, M., and Lorenz, W. 2018. 'The Metal Roof Truss of the Pantheon's Portico in Rome,' International Journal of The Construction History Society, 33.2, 1–22, 2018

Michler, A., *Hyperlocalization of Architecture*. Los Angeles: eVolo Press, 2015

Potter, C., *Welcome to the Circular Economy: The Next Step in Sustainable Living*. London: Laurence King, 2021

Robbin, T., *Engineering a New Architecture*. Newhaven: Yale University Press, 1996

Ruby, Andreas and Ilka (et al), *Re-inventing Construction*. Berlin: Ruby Press, 2010

Solanki, S., *Why Materials Matter: Responsible Design for a Better World*. Munich: Prestel, 2018

Stanwix, W. and Alex Sparrow, *The Hempcrete Book: Designing and Building with Hemp-lime*. London: Green Books, 2023

Stauffer, Nancy W., 'Using nature's structures in wooden buildings'. Autumn 2021 issue of *Energy Futures*, MIT

Van Aubel, Marjan, *Solar Futures: How to Design a Post-Fossil World with the Sun*. Prinsenbeek: Jap Sam Books, 2022

Vincent, J., *Structural Biomaterials*. New Jersey: Princeton University Press, 1990

Zhang, H. (et al), *A new concept of bio-based prestress technology with experimental Proof-of-Concept on Bamboo-Timber composite beams. Construction and Building Materials*, Volume 402, 26 October 2023, Elsevier, 132991

WEBSITES

Block Research Group, Institute of Technology in Architecture at ETH Zürich, research into structural design solutions: www.block.arch.ethz.ch/brg/research

Circular Ecology, embodied carbon database for building materials: bit.ly/4dd3b0s

Circular Economy, Ellen MacArthur Foundation: www.ellenmacarthurfoundation.org

Gnanli Landrou, clean concrete interview: www.bbc.co.uk/programmes/w3ct3j2b

The Institution of Structural Engineers, how to calculate embodied carbon: bit.ly/3VJ7Wrp

International EPD System, established with a Life Cycle Assessment: www.environdec.com/home

LCA Building Life Cycle Stages diagram: bit.ly/3xCyS3Z

Material District, materials database for innovative materials: materialdistrict.com

Resource Efficiency Collective (University of Cambridge): www.reficciency.org

Use Less Group, Dr Julian Allwood (University of Cambridge): www.uselessgroup.org/research/construction

Whole Life Carbon Assessment (WLCA) for the built environment: Royal Institution of Chartered Surveyors (RICS): bit.ly/3zcuskl

Will Stanwix, hempcrete website: thathempcreteguy.com

Glossary

Bagasse: Dry waste pulp fibre left over from processing sugarcane and other plants including sisal, olives and grapes. It can be burnt as a fuel or processed for its cellulose for making paper, board and other products.

Biogenic Carbon: Biogenic carbon is the carbon that is stored or sequestered in a given material. While growing, trees sequester and store carbon which can be accounted for as part of a building's Life Cycle Analysis (LCA).

BREEAM: Building Research Establishment Environmental Assessment Method. Developed by the Building Research Establishment in 1990, BREEAM is an assessment and certification method for measuring a building's sustainability, based on material specification, design parameters, construction and building operation.

The British Board of Agrément (BBA): The BBA is the UK's largest independent certification body for construction products and systems.

Carbon Sequestration: The capture and storage of carbon dioxide (CO_2) from the Earth's atmosphere. During its growth, timber naturally sequesters carbon dioxide.

Circular Economy: This is where materials never become waste and nature is regenerated; products and materials are kept in circulation through processes like maintenance, reuse, refurbishment, remanufacture, recycling and composting. See the Ellen MacArthur Foundation website.

Cob: Cob is an earth-based natural material made from a mixture of subsoil, organic fibre such as straw, and water. Lime, sand and clay are sometimes added for binding and finish.

Cross-laminated Timber (CLT): Laminate of three or more layers of solid timber, laid in perpendicular layers bonded with structural adhesives and used for structural walls and floors. The timber is typically from a softwood such as spruce or larch.

Design for Disassembly (DfD): Design for buildings and structures that can be successfully disassembled without inherently compromising the component parts and thus minimizing waste. An example is to use a bolted connection rather than a glued or bonded joint.

Embodied Carbon: Embodied or embedded carbon is the carbon dioxide emitted when producing materials and products. It is a total of the energy used to extract and transport raw materials in addition to emissions associated with the manufacturing process. The embodied carbon of a material is also described as the Global Warming Potential (GWP). Embodied Carbon or GWP is measured in CO_2e/kg.

Environmental Product Declaration (EPD): An EPD is the environmental impact measurement of a product calculated using a Life Cycle Assessment measuring carbon emissions in the sourcing, manufacture, transportation and end-of-life use including recycling potential. An EPD can be sourced from individual material suppliers or online EPD databases. While there are European and International standards for an EPD, not all products have them, especially from smaller suppliers.

Faience: Manufacture of slip-cast fired ceramic components in a centuries-old process imported from Italy. The liquid slip-cast process allows for the casting of complex forms.

Feedstock: A bulk material used in the industrial production of materials.

Global Warming Potential (GWP): See Embodied Carbon.

Glue Laminated Timber (Glulam): The process of creating engineered timber members such as beams and columns from a lamination of thinner timber members bonded together with adhesive. The grain of the laminations runs the length of the timber member (as opposed to CLT). Glulam typically uses softwood timber such as spruce or Douglas fir and the process allows for the forming of large-scale (and curved) members for domes and bridges.

Glulam technology has been successfully used for over a century.

Laminated Veneer Lumber (LVL): A laminate of thin timber sheet veneers (typically 3mm [⅛ in] each layer) laid in parallel layers bonded with structural adhesives. LVL is stronger than sawn timber sections and CLT because any imperfections such as knots are limited to the single veneer, and as such the combined laminate is very dense and stiff. LVL is made from softwood such as spruce as well as hardwood such as beech.

Life Cycle Assessment (LCA): The LCA for a building is an analysis of the environmental impact of all the stages of a building's life. It considers the raw materials and energy used in production, construction, operation, maintenance and end-of-life disposal. Building life cycle information is classified as follows:

A1–3 Product stage: (A1) Raw materials supply; (A2) Transport; (A3) Manufacturing.

A4–5 Construction process stage: (A4) Transport; (A5) Construction installation process.

B1–7 Use stage: (B1) Use; (B2) Maintenance; (B3) Repair; (B4) Replacement; (B5) Refurbishment; (B6) Operational energy use; (B7) Operational water use.

C1–4 End of life stage: (C1) De-construction or demolition; (C2) Transport; (C3) Waste processing; (C4) Disposal.

D Reuse, recovery and/or recycling potentials; expressed as net impacts and benefits.

Low Energy Transformation Initiative (LETI): LETI is a voluntary network, established in 2017 of over 1000 built-environment professionals, with the aim of putting the UK and the planet on the path to a zero-carbon future.

Operational Energy or Operational Carbon: The emissions resulting from energy and water consumption arising from the use of technical systems in the building over its life. These emissions include all heating, cooling, lighting, and hot and cold water supply.

Passivhaus: An energy performance standard designed to minimize space heating requirements with high levels of thermal insulation.

Pozzolana: The use of volcanic ash that was first used by the ancient Greeks and Romans as a binder for a concrete-like substance. Pozzolana, which has been superheated through the volcanic process, has cementitious properties and is used as a lower-carbon cement replacement material.

Recycling: The process of converting waste materials into new materials and products. It is important to distinguish between a recycling process such as crushing, remelting and forming glass bottles, and recycling where glass bottles might be simply cleaned and reused, saving considerable energy. This process can be applied to the construction industry with the successful reuse or re-purposing of building materials and components.

Reuse: Also referred to as upcycling. See Recycling.

Thermoplastic: A thermoplastic is a polymer that is softened with heat and can be extruded and moulded. This process is endlessly repeatable.

U-value: The U-value is the measure of thermal transmittance of a given material. Used to measure thermal insulation products, the U-value is measured in W/m^2 K. The lower a U-value, the better the thermal insulative value of a material is. An R-value is the measure of thermal resistance and the reciprocal of a U-value, and as such the higher an R number, the better the thermal insulative value.

Waste Stream: A waste stream is the flow of waste from its source to its end. This destination can be the recovery/reuse, recycling or disposal of the waste.

Index

Page numbers in *italics* indicate illustration captions.

15 Clerkenwell 85

3D printing technology 29
  3D-printed earth house (TECLA) 48–9
  3D-printed mycelium 117, 176–7
  3D-printed reefs 29, 80–1
  House of Cores, Houston, Texas 29, 70–3
  Sandwaves, The 29, 111–12
  TerraCool 45

Abeille, Joseph 162
Accoya 119, 120, *136*, 137
Air Bubble *183*, 183–4, *184*
AKT II 46–7
algae 180
  air-purifying eco-machine 183–4
  bio-cement blocks from microalgae 117, *181*, 181–2, *182*
Ally Capellino 155
Alsop Architects 227
aluminium 29
Amorim 137
Andersson, Liv 74–5, *75*
Architects' Journal 57
architecture 8–11
ARUP 99–101
Ashen House, Ithaca, New York 70
Atelier LUMA 117, 199–201

babassu coconuts *166*, 166–7, *167*
Babic, Eva 101
bagasse 154, *155*, 162
Balbaligo, Jan *25*, 117, 156–8
bamboo 117, 150
  bamboo & timber composite gridshell dome structure 117, *159*, 159–61, *161*
  bamboo construction *25*, *156*, 156–8, *158*
Ban, Shigeru *20*, 148
Banfield, Simon 62
BanfieldWood 62–5
Barkow Leibinger 90–2
Bartlett School of Architecture, University College London 68–9, 137

Basnett, Pooja 192–3
BC Materials 56, *57*, 57–8, 199
Bell, Bruce 131–2
Bennetts Associates 25, 34–5
Bini, Dante 29, 66–7
Bio-block™ Spiral *181*, 181–2, *182*
biopolymers 186
  bacterial biopolymers 192–3, *193*
  shellfish bioplastic 117, 190–1
bioreceptive living walls 68–9
Biozeroc *74*, 74–5
Black & White Building, London *121*, 121–3, *123*
Blast Studio 117, 176–7
Block, Philippe 117, 178
Bonwetsch, Tobias 28, 36
Braungart, Michael *Cradle to Cradle: Remaking the Way We Make Things* 24
Brewin, Peter 79
'Brick for Venice, A' *28*, 46–7
Bristogianna, Telesilla 29, 104–7
Buckminster Fuller, Richard *22*, 24
Burney, Henna 117, 199
Burrell Renaissance Project, Glasgow 29, 99–101

Caen University 38–9
Candilis, Josic, and Woods 87
CarbiCrete *51*
carbon (ac)counting 21–2
Carmody Groarke 29, 56–8
CCSC (Carbonatable Calcium Silicate Clinkers) 53
cement 17, 28, 29, 31–2, 79
  bio-cement blocks from microalgae 117, *181*, 181–2, *182*
  cement-free concrete 54–5, 74–5
  cement-free construction 34–5, 37, 40–1, *51*, 61
  poikilohydric (bioreceptive) living walls 68–9
  supplementary cementitious material (SCM) 51–3
Cemex 73
Chapman, Samuel 60
circular economy 24–5
CIVE 73
clay 28
  CobBauge 38–9
  slip-cast clay faience 42–5
  straw & clay compressed earth blocks 34–5

Cleancrete 54–5
CNC machining *43*, 97, 131, *132*, 137, 141, 144
CobBauge 38–9
COBOD construction printer 73
Coimbra University, Portugal 69
concrete 50
  3D-printed concrete 29, 70–3
  cement-free concrete 54–5, 74–5
  Essential Homes Research Project 76–9
  fabric formwork 29, 66–7
  Foredown recycled tiles 62–5, *65*
  recycled construction waste bricks 60–1
  recycled crushed concrete & glass bricks 56–8
  supplementary cementitious material (SCM) 51–3
  trabeculated panels 80–1
  zero-carbon concrete *74*, 74–5
Concrete Canvas 29, 79
construction 8–11
  history of construction 14–17
cooling systems 42–5, *45*
Cooper, Oscar 154–5
cork 69, 116
  Cork House, Berkshire 116, *136*, 136–7
Crawford, William 79
cross-laminated timber (CLT) 37, 94, 116, 121, 129, 135, *137*
Cruz, Marcos 68–9
Current Window 29, *102*, 102–3

D-Shape 80–1
Darmstadt Technical University 117, 148–9
DeBrincat, Graeme 101
Delft University of Technology 104–5
Deme, Doudou 31
Demoulin, Thibault 55
Design Museum, Gent 56–8, *58*
Dini, Enrico 29, 80–1
DOSU Studio 29, 96–7
Düsing, Gustav 29, 94–5

earth 28, 30
  3D-printed earth house (TECLA) 48–9
  CobBauge 38–9
  earth bricks & typha reeds, Dakar 31–3
  earth-timber floor slab 36–7
  rammed earth 40–1
  silt brick 46–7
  straw & clay compressed earth blocks 34–5

Earth Building UK and Ireland 38–9, 41
East London University 117, 162–4
ecofitted VW Golf *19*
ecoLogicStudio 117, 183–4
Edinburgh Napier University 117, 159
eelgrass *117*
  Modern Seaweed House, Læsø,
    Denmark *169*, 169–70, *170*
Elbphilharmonie, Hamburg *11*, 117,
  141, 141–2, *142*
Elementerre 31–3
Engel, Heino Structure Systems 90
Environmental Product Declaration
  (EPD) 61
'Equanimity' stone beam 85, *85*
Essential Homes Research Project 76–9

fabric formwork 29, 66–7
Facit Homes 116, *131*, 131–2, *132*
Fahy, Lachlan 42–5
fashion 18–19
Fiction Factory 117, 144–7
Flat House, Cambridgeshire *154*, 154–5,
  155
flax 129, *144*, 151
  hemp & flax fibre house 154–5
fly ash 53, 74
Foredown recycled tiles 62–5, *65*
Forest Stewardship Council (FSC) 116, 131
Formby Stair *84*, 85
Foster + Partners 29
furan 110, 111–12

Gehry, Frank 199–201
General Electric 73
Gent Waste Brick 29, 56–8, *57*, *58*
GGBS (ground granulated blast-furnace
  slag) 53, 74
Glasgow Botanic Gardens *17*
glass 29, 98
  cullet 101
  photovoltaic solar energy 102–3
  recycled crushed concrete & glass bricks
    56–8
  recycled glass bricks 104–7
  recycling construction glass 99–101
  Water-filled Glass (WFG) 108–9
Glass & Transparency Research Group,
  Delft University of Technology 104–5
Gomes, Paulo 117, 187
Graduate School of Construction Engineers
  of Caen (ESITC) 38–9

grasses 117, 150
  sugarcane waste construction blocks
    117, *162*, 162–4, *164*
Grimshaw Architects 117, 162–3
grown materials 116–17
Guanxi University of Science and
  Technology 159
Gutai, Matyas 108–9
gypsum 141–2

H.G. Matthews 25, 34–5
Habert, Guillaume 54–5
Hacke, Max 29, 94–5
HANNAH 29, 70–3
Hebel, Dirk 117, 178
hemp 18, 117, 150
  CobBauge 38–9
  hemp & flax fibre house *154*, 154–5
  hempcrete (hemp lime composite
    building material 117, *151*, 151–3, *153*
Herzog & de Meuron *11*, 28, 36, 37, 117,
  141–2
HiLo Lab, University of British Columbia
  117, 139–40
Holcim 73, 75, 76–9
HORTUS building, Basel 37
House of Cores, Houston, Texas 29, 70–3
Howland, Matthew Barnett 136–7
Hudson Architects 38–9
Hydrox Wall 117, *139*, 139–40, *140*
hyperlocal sourcing 40, 57, 128
  hyper-local architecture 166–7

Inner Mongolia University of Science
  and Technology 159
InVert™ 29, 97

John McAslan & Partners 99–101

K-Briq® 29, *60*, 60–1, *61*
Kane, Fred L. 141
Karlsruhe Institute of Technology (KIT),
  Germany 178
Keable, Rowland 28, 40–1
Kenoteq 60–1
Knauf 141

laminated veneer lumber (LVL) 121, 135
  timber gridshell in laminated veneer
    lumber (LVL) 124–6
Landrou, Gnanli 54–5
Le Corbusier 87

Life Cycle Analyses/Assessments (LCAs)
  24, 122
  Strocks® *34*, 35, *35*
Local Works Studio, Sussex *28*, 46–7, 57–8
LoDo Lab, University of British Columbia
  139–40
Lok, Leslie 70
Luma Foundation, Arles 117, 199

Maisons Tropicales *87*, 87–8, *88*
Mamou-Mani Architects 29, 111–12
Mamou-Mani Arthur 111
Margent Farm, Cambridgeshire 154–5
Mario Cucinella Architects (MCA) 48–9
materials 8–11
  history of construction materials 14–17
  material properties 11
Mausbach, Artur 19–20
Mboup, Nzinga *28*, 31–3
McDonough, William *Cradle to Cradle:
  Remaking the Way We Make Things* 24
McKevitt, Steve *The Solar Revolution* 102
Medero, Gabriela 60
metals 86
  lightweight shell construction 90–2, *92*
  Maisons Tropicales *87*, 87–8, *88*
  Study Pavilion TU Braunschweig,
    Germany 29, 94–5, *95*
  thermobimetal 29, *96*, 96–7, *97*
Michels, Neil 57
Michler, Andrew *The Hyperlocalization of
  Architecture* 167
MIT Building Technology Program 134–5
Modern Seaweed House, Læsø, Denmark
  169, 169–70, *170*
Mogu 117, 173–4
MOMS (Magnesium Oxides from
  Magnesium Silicates) 53
Montreal Biosphere *22*
Moretti, Massimo 48
Morgante, Andrea 80
Mueller, Caitlin 134–5
Multihalle, Mannheim 124
mycelium 117, 172
  3D-printed mycelium 117, 176–7
  mycelium construction materials 117,
    173, 173–4, *174*
  MycoTree 117, *178*, 178–9, *179*

Nagata Acoustics 141
Nielsen, Søren 170
Niemeyer, Oscar 87

Nieuwe Veemarkt, De, Netherlands 25,
    116, 128–30, *130*
Nieva, Germa 166–7
Norman Foster Foundation 76–9

Oikonomopoulu, Faidra 29, 104–7
Oscarson, Elding 116, 124–6
Otto, Frei 124
Oxara 54–5, *55*

palm 150
    building with babassu *166*, 166–7, *167*
paper 117, 138
    cardboard house 144–7
    circular exterior wall system made of
        paper-based materials 117, 148–9
    paper and gypsum 'white skin' 117,
        141–2
    paper pulp concrete *139*, 139–40, *140*
Paper Dome *20*
PERI 73
Peterborough Cathedral, UK *14*
PFA (pulverized fuel ash) 53
photovoltaic solar energy 102–3
pisé de terre (pisé construction) 40
plasterboard (drywall) 141–2, 148
Plymouth University 38–9
poikilohydric (bioreceptive) living
        walls 68–9
Portland cement 17, 51, 53
post-tensioned stone 29, 83–5
pozzolans 53
Precht, Chris 111
Prometheus Materials 117, 181–2
Prouvé, Henri 87
Prouvé, Jean 29, 87–8

rammed earth 40–1
    formwork 41, *41*
Re3 29, 104–7
recycled polymers 186
    recycled plastic waste 117, 187–8, *188*
reduce, reuse, recycle and repeat 24–5
Rematter 28, 36–7, 129
Rice, Peter 83–5
Ricketts, Sian 57
Ross, Charles 18
Ryan, Tony *The Solar Revolution* 102

Sackett, Augustine 141
Sajid, Mohsin 18
salt 198

wall of salt and salt door handles *199*,
        199–201, *201*
sand 110
    The Sandwaves 29, 111–12
Sayed, Ehab 20
Schiano-Phan, Rosa 42
Schilling, Oep 147
Schlaich Bergermann 29, 90–2
Schwartz, Hermann 42
seaweed 168
Shahpazov, Nikolay 25, 28, 34–5
Sharp Centre for Design, Toronto *22*
sheep's wool 196–7
Shellmet 117, *190*, 190–1, *191*
Sibbel, Karljin 117, 199
Skidmore, Owings & Merrill (SOM) 181, 182
Snel, René 144
Sparrow, Alex *The Hempcrete Book* 151–3
SQIM 117, 173–4
Stanwix, William 117, 154–5
    The Hempcrete Book 151–3
steel 29, 86
stone 29, 82
    post-tensioned stone 29, 83–5
Stonemasonry Company, The 83, 85
Stora Enso 124, 125
straw 28
    straw & clay compressed earth blocks
        34–5
Strocks®, London *34*, 34–5
Studio Nauta 25, 116, 128–30
Studio Precht 29, 111–12
Study Pavilion TU Braunschweig, Germany
        29, 94–5, *95*
Sugarcrete® *162*, 162–4, *164*
Sung, Doris 29, 96–7
supplementary cementitious material
        (SCM) 51–3
Swiss Federal Institute of Technology
        (ETH), Zürich 178

Taha, Amin 29, 83, 85
Tate & Lyle Sugars 162–3
TBWA\HAKUHODO + Quantum 117, 190–1
technology transfer 18–19
    product design 19–20
TECLA (Technology and Clay) 48–9
Temel, Dilara 42–5
TerraCool 11, 42–5
textiles 18–19
thermobimetal 29, *96*, 96–7, *97*
timber *8*, 28, 116, 118

bamboo & timber composite gridshelll
        dome structure 117, 159–61
Black & White Building, London *121*,
        121–3, *123*
digitally manufactured timber house
        131–2
earth-timber floor slab 36–7
Nieuwe Veemarkt, De 25, 116, 128–30,
        130
non-chemical timber treatments 119–20
timber 'tree fork' connections 134–5, *135*
timber gridshell in laminated veneer
        lumber (LVL) 124–6
Torres, Manel 18–19
Toyota, Yasuhisa 141
trabeculated panels 80–1
Transformation Workshop, Brazil 117,
        187–8, *188*
Tree Column *176*, 176–7, *177*
Trumpf Footbridge, Ditzingen, Germany 29,
        90–2, *92*
typha reeds 33
    earth bricks & typha reeds, Dakar 31–3

UN: *Bend the trend: Pathways to a liveable
    planet as resource use spikes* 28
UNESCO 40
Urban Radicals *28*, 28, 46–7

Van Aubel, Marjan 29, 102–3
Vandkunsten Architects 169–70
Vastern Timber 119, 120
Venice Biennale 2023 46–7, 76, *79*

Water-filled Glass (WFG) 108–9
Waugh Thistleton Architects 116, 121–2
Webb Yates 83–5
Webb, Steve 25, 29, 83–5
Wikkelhouse 117, 132, *144*, 144–7, *147*
Wilton, Oliver 136–7
Wisdome, Stockholm 116, 124–6, *126*
wood-and-loam floors 36
Wood, Matthew 62
wool 194
    insulation & composite 195–7, *197*
World's Advanced Saving Project (WASP)
        48–9
Worofila 31–3

zero-carbon concrete *74*, 74–5
Zhang, Hexin (Johnson) 117, 159–61
Zivkovic, Sasa 70

Credits

THE AUTHORS WOULD LIKE TO THANK

Samantha Hardingham, Bert McLean, Freda McLean, Flora McLean, Scott Batty, Francois Girardin, Chris Leung, Teo Cruz, Nzinga Mboup, Nikolay Shahpazov, Tobias Bonwetsch, Steve Goodhew, Karen Hood-Cree, Rowland Keable, Lachlan Fahy, Dilara Temel, Era Savvides, Nasios Varnavas, Laura Iloniemi, Gnanli Landrou, Sian Ricketts, Lucy Black, Gabriela Medero, Dante Bini, Marcos Cruz, Leslie Lok, Sasa Zivkovic, Liv Andersson, Santiago Riveiro, Dino Enrico, Steve Webb, Doris Sung, Mike Schlaich, Gustav Düsing, Max Hacke, Marjan van Aubel, Faidra Oikonomopoulou, Telesilla Bristogianni, Matyas Gutai, Arthur Mamou-Mani, Kirsten Haggart, Elding Oscarson, Jan Nauta, Benjamin Filbey, Bruce Bell, Matthew Barnett Howland, Dido Milne, Oliver Wilton, Dorus Stolk, William Stanwix, George Massoud, Paloma Gormley, Jan Balbaligo, Hexin (Johnson) Zhang, Armor Gutiérrez Rivas, Alan Chandler, Michelle Lange, Nausicaa Foglia, Paola Garnousset, Philippe Block, Shalika Oza, Maria Azzurra Rossi, Paulo Gomes, Pooja Basnett, Karlijn Sibbel, Henna Burney, Adam Thwaites, Sofia Steffenoni, German Nieva, Simon Banfield, Tom Hesslenberg, Ana Gatóo, Urna Sodnomjamts.